Katharina Lückerath

Fas Ligand reverse signaling: an immunmodulatory setscrew

Katharina Lückerath

Fas Ligand reverse signaling: an immunmodulatory setscrew

Insights gained from 'knockout/knockin' mice that express a functional Fas Ligand molecule lacking the intracellular domain

Südwestdeutscher Verlag für Hochschulschriften

Imprint

Any brand names and product names mentioned in this book are subject to trademark, brand or patent protection and are trademarks or registered trademarks of their respective holders. The use of brand names, product names, common names, trade names, product descriptions etc. even without a particular marking in this work is in no way to be construed to mean that such names may be regarded as unrestricted in respect of trademark and brand protection legislation and could thus be used by anyone.

Publisher:
Südwestdeutscher Verlag für Hochschulschriften
is a trademark of
Dodo Books Indian Ocean Ltd., member of the OmniScriptum S.R.L Publishing group
str. A.Russo 15, of. 61, Chisinau-2068, Republic of Moldova Europe
Printed at: see last page
ISBN: 978-3-8381-2487-2

Zugl. / Approved by: Frankfurt (Main), Goethe-Universität Frankfurt, Diss., 2010

Copyright © Katharina Lückerath
Copyright © 2011 Dodo Books Indian Ocean Ltd., member of the OmniScriptum S.R.L Publishing group

Table of contents

Summary 7

1 Introduction 11

1.1 Apoptosis 11

   1.1.1 The extrinsic pathway of apoptosis 12

   1.1.2 The intrinsic pathway of apoptosis 13

1.2 The FasL/Fas system 14

   1.2.1 Molecular mechanism of FasL/Fas-induced apoptosis 14

   1.2.2 Apoptotic Fas signaling in the immune system 15

      1.2.2.1 Lymphocyte death in the periphery: activation-induced cell death 16

      (AICD) and activated cell autonomous cell death (ACAD) 16

      1.2.2.2 FasL/Fas-induced apoptosis in B cell function 17

   1.2.3 Apoptotic Fas signaling outside the immune system 18

   1.2.4 FasL/Fas signaling in the establishment of immune privilege and in 18

   tumor biology 18

   1.2.5 Non-apoptotic Fas signaling 20

1.3 The Fas Ligand 22

   1.3.1 FasL structure 22

   1.3.2 Regulation of FasL expression and activity 23

      1.3.2.1 Transcriptional control 23

      1.3.2.2 FasL sorting and storage 23

      1.3.2.3 Regulation of FasL activity by dynamic localization in lipid rafts 24

      1.3.2.4 FasL processing 25

      1.3.2.5 Regulation of FasL by proteins that bind to its intracellular domain 26

1.4 Reverse ligand signaling 27

   1.4.1 Reverse signal-transduction by TNF family ligands 27

   1.4.2 FasL reverse signaling 28

   1.4.3 Functional implications of FasL reverse signaling 29

1.5 Aims of the project 31

2 Materials and methods 32

2.1 Materials 32
2.1.1 Equipment 32
2.1.2 Consumables 33
2.1.3 Chemical reagents 33
2.1.4 Inhibitors 35
2.1.5 Enzymes 35
2.1.6 Size standards 35
2.1.7 Commercial kits 35
2.1.8 Buffer and solutions 36
2.1.9 Antibodies 37
2.1.10 Vectors 39
2.1.11 Oligonucleotides 39
2.1.12 Mouse lines 40

2.2 Methods 41
2.2.1 Animal models 41
2.2.2 Genotyping 41
2.2.3 Cell culture methods 42
2.2.3.1 Isolation of naïve murine lymphocytes 42
2.2.3.2 Generation of T cell blasts 42
2.2.3.3 Lymphocyte maintenance and stimulation 43
2.2.3.4 Erythrocyte lysis 43
2.2.3.5 Determination of cell number and viability 43
2.2.3.6 Cultivation of cell lines 44
2.2.3.7 Freezing and thawing of cells 44
2.2.3.8 Transfection of cells 44
2.2.4 Flow cytometry 45
2.2.4.1 Analysis of cell surface marker expression 45
2.2.4.2 Intracellular stainings 45
2.2.4.3 Staining of unfixed cells with Annexin V and propidium iodide 46
2.2.4.4 Propidium iodide staining of ethanol fixed cells – Nicoletti test 46
2.2.4.5 Carboxyfluorescein diacetate succinimidyl ester (CFSE) dilution assay for lymphocyte proliferation 46
2.2.4.6 Co-culture of primary lymphocytes with Fas-sensitive A20 target cells 47
2.2.4.7 β-Galactosidase activity assay for the analysis of Wnt signaling *in vivo* 47

2.2.5 Enzyme-linked immunoabsorbant assay (ELISA) 48
   2.2.5.1 Cell-based ELISA 48
   2.2.5.2 Standard solid phase sandwich ELISA 50

2.2.6 Molecular biological methods 51
   2.2.6.1 Isolation of total RNA 51
   2.2.6.2 Quantification of nucleic acid concentration and purity 51
   2.2.6.3 Complementary DNA (cDNA) synthesis 51
   2.2.6.4 Polymerase chain reaction (PCR) 52
   2.2.6.5 Electrophoresis of PCR products 52
   2.2.6.6 Quantitative real time polymerase chain reaction (qRT-PCR) 52
   2.2.6.7 mRNA expression profiling of Wnt signaling-related genes 54

2.2.7 Protein-biochemical methods 55
   2.2.7.1 Protein-extraction from mammalian cells 55
   2.2.7.2 Determination of protein concentration using the Bradford method 55
   2.2.7.3 SDS polyacrylamide gel electrophoresis (PAGE) 55
   2.2.7.4 Immunoblotting 56

2.2.8 In vivo studies 56
   2.2.8.1 Analysis of thymocyte proliferation - Bromodesoxyuridine (BrdU) 56
   incorporation 56
   2.2.8.2 Expansion of Vβ8 T cell receptor chain-expressing T cells 57
   2.2.8.3 Germinal center formation 57
   2.2.8.4 Thioglycollate-induced peritonitis 60
   2.2.8.5 Statistics 61

3 Results 62

3.1 Mouse model for FasL reverse signaling deficiency 62

3.2 *FasL ΔIntra* mice represent a suitable model to study FasL reverse signaling 63
   3.2.1 FasL ΔIntra mice express functional FasL that is capabale of inducing 63
   apoptosis 63
   3.2.2 FasL ΔIntra mice do not display obvious phenotypic anomalies 66

3.3 Analysis of FasL reverse signaling *ex vivo* 68
   3.3.1 The FasL ICD impairs lymphocyte proliferation 68
   3.3.2 Defective FasL reverse signaling accounts for the observed enhanced 70
   proliferation of FasL ΔIntra lymphocytes 70
      3.3.2.1 Comparable extent of cell death in *FasL ΔIntra* and wildtype lymphocytes 70

    3.3.2.2 Non-apoptotic signaling through Fas does not account for the proliferative differences in *FasL ΔIntra* and wildtype lymphocytes 71

  3.3.3 The FasL ICD regulates ERK1/2 activation and proliferation by influencing 72

  phosphorylation of PLCγ2 and PKC 72

    3.3.3.1 The presence of the FasL ICD reduces ERK1/2 activation 72

    3.3.3.2 MAP-kinases upstream of ERK1/2 do not appear to be regulated by FasL 74 reverse signaling 74

    3.3.3.3 FasL reverse signaling *via* PLCγ2 and PKC regulates ERK1/2 activation .. 75 and proliferation 75

    3.3.3.4 The PI3K/Akt pathway is apparently not involved in FasL reverse signaling 77

  3.3.4 Identification of FasL reverse signaling target genes 78

    3.3.4.1 Global gene expression profiling 78

    3.3.4.2 FasL reverse signaling regulates genes associated with lymphocyte .... 78 proliferation and activation 78

    3.3.4.3 Significant regulation of Wnt signaling pathway-associated genes by the .. 80 FasL ICD 80

  3.3.5 B cells isolated fom spleen express Lef-1 81

3.4 *In vivo* studies to investigate the consequences of FasL reverse signaling 82

  3.4.1 Analysis of thymocyte proliferation 82

  3.4.2 Expansion of Vβ8 T cell receptor (TCR) chain-expressing T cells 84

  3.4.3 Anti-viral response of $CD8^+$ T cells following Lymphocytic choriomeningitis 85

  virus (LCMV) infection 85

  3.4.4 Acute and long-term immunity in an infection model of listeriosis 86

  3.4.5 Participation of the FasL ICD in Ovalbumin-induced allergic airway disease 90

  3.4.6 FasL reverse signaling modulates the germinal center reaction 92

  3.4.7 Thioglycollate-induced peritonitis as a model for neutrophil migration 95

4 Discussion 96

  4.1 FasL ΔIntra mice represent a suitable model for FasL reverse signaling .... 96 deficiency and do not display an abnormal phenotype 96

  4.2 Signaling via the FasL ICD impairs activation-induced lymphocyte .... 98 proliferation by a mechanism involving PLCγ2, PKCα and ERK1/2 98

  4.3 Reverse FasL signals regulate target gene transcription: differential .... 101 expression of pro-proliferative and Lef-1-dependent genes in FasL ΔIntra mice......101

4.4 Molecular model for FasL reverse signaling 102

4.5 Analysis of FasL reverse signaling in vivo 106
    4.5.1 Signaling via the FasL ICD is apparently not important for thymocyte 106
    development 106
    4.5.2 FasL reverse signaling modulates the germinal center reaction 107
    4.5.3 Signals transmitted through the FasL ICD participate in Ovalbumin-induced 108
    allergic airway disease...108
    4.5.4 Regulatory mechanism operating in vivo mask FasL reverse signaling.......109

4.6 Conclusion..110

5 Literature 111

6 Appendix 121
    6.1 Systematic expression analysis of T cells..121

    6.2 Abbreviations..129

Summary

Fas Ligand (FasL; CD95L; CD178; TNSF6) is a 40 kDa glycosylated type II transmembrane protein with 279 aa in mice and 281 aa in humans that belongs to the tumor necrosis factor (TNF) family. The extracellular domain (ECD) harbors a TNF homology domain, the receptor binding site, a motif for self assembly and trimerization, and several putative N-glycosylation and a metalloprotease cleavage site/s. The cytoplasmic tail of FasL is the longest of all TNFL family members and contains several conserved signaling motifs, such as a putative tandem Casein kinase I phosphorylation site, a unique proline-rich domain (PRD) and phosphorylatable tyrosine residues (Y7 in mice; Y7, Y9, Y13 in human).

The FasL/Fas system is renowned for the potent induction of apoptosis in the receptor-bearing cell and is especially important for immune system functions. It is involved in the killing of target cells by natural killer (NK) and cytotoxic T cells, in the (self) elimination of effector cells following the proliferative phase of an immune response (activation-induced cell death; AICD), in the maintenance of immune-privileged sites and in the induction and maintenance of peripheral tolerance. Owing to its potent pro-apoptotic signaling capacity and important functions, FasL expression and activity are tightly regulated at transcriptional and post-transcriptional levels and restricted to few cell types, such as immune effector cells and cells of immune-privileged sites. In contrast, Fas is expressed in a variety of tissues including lymphoid tissues, liver, heart, kidney, pancreas, brain and ovary.

In addition to its pro-apoptotic function, the FasL/Fas system can also elicit non-apoptotic signals in the receptor-expressing cell. Among others, Fas-signaling exerts co-stimulatory functions in the immune system, e.g. by promoting survival, activation and proliferation of T cells.

Besides the capacity to deliver a signal into receptor-bearing cells ('forward signal'), FasL can receive and transmit signals into the ligand-expressing cell. This phenomenon has been described for several TNF family ligands and is known as 'reverse signaling'. The first evidence for the existence of reverse signaling into FasL-bearing cells stems from two studies that demonstrated either co-stimulation of murine $CD8^+$ T cell lines by FasL cross-linking or inhibition of activation-induced proliferation of murine $CD4^+$ T cells. In both cases, the observed changes of proliferative behaviour critically depended on the presence of a signaling-competent FasL. Almost certainly, the FasL ICD is functionally involved in signal-transmission: (i) The ICD is highly conserved across species and harbors several signaling motifs, most notably a unique PRD. (ii) Numerous proteins have been identified which interact with the FasL PRD *via* their SH3 or WW domains and regulate various aspects of FasL biology, such as FasL sorting, storage, cell surface expression and the linkage of FasL to intracellular signaling pathways. (iii) Post-translational modifications of the ICD have been implicated in the sorting of FasL to vesicles and the FasL-dependent activation of Nuclear factor of activated T cells (NFAT). (iv) Proteolytic processing of FasL liberates the ICD and allows its translocation into the nucleus where it might influence gene transcription. (v) It could be shown that overexpression of the FasL ICD is sufficient to initiate reverse signaling upon concomitant T cell receptor (TCR) stimulation and ICD cross-linking.

Conflicting data on the consequences of FasL reverse signaling exist, and co-stimulatory as well as inhibitory functions have been reported. These discrepancies probably reflect the use of artificial experimental systems. Neither the precise molecular mechanism underlying FasL reverse signaling, nor its physiological relevance have been addressed at the endogenous protein level *in vivo*.

Therefore, a 'knockout/knockin' mouse model in which wildtype FasL was replaced with a deletion mutant lacking the intracellular portion (*FasL ΔIntra*) was established in the group of PD Dr. Martin Zörnig. In the present study, *FasL ΔIntra* mice were phenotypically characterized and were employed to investigate the physiological consequences of FasL reverse signaling at the molecular and cellular level.

To ensure that *FasL ΔIntra* mice represent a suitable model to study the consequences of FasL reverse signaling, we demonstrated that activated lymphocytes from homozygous *FasL ΔIntra* or wildtype mice express comparable amounts of (truncated) FasL at the cell surface. The truncated protein retains the capacity to induce apoptosis in Fas receptor-positive target cells, as co-culture assays with FasL-expressing activated lymphocytes and Fas-sensitive target cells showed. Additionally, systematic screening of unchallenged mice did not reveal any phenotypic abnormalities. Notably, signs of a lymphoproliferative autoimmune disease associated with FasL-deficiency could not be detected.

As several reports have implicated FasL reverse signaling in the regulation of T cell expansion and activation, proliferation of lymphocytes isolated from *FasL ΔIntra* and wildtype mice in response to antigen receptor stimulation was investigated. Using CFSE dilution assays it could be demonstrated that the proliferative response of $CD4^+$ T cells, $CD8^+$ T cells and of B cells was enhanced in the absence of the FasL ICD. Interestingly, this effect was most pronounced in B cells and could only be detected in $CD4^+$ T cells after depletion of $CD4^+CD25^+$ regulatory T cells. To our knowledge, this is the first time that FasL reverse signaling has been demonstrated in B cells.

In a series of experiments, the activation of several pathways that are known to play important roles in signal-transmission initiated upon antigen receptor triggering was assessed. As a molecular correlate for the observed enhancement of activation-induced proliferation, Extracellular signal regulated kinase (ERK1/2) phosphorylation was significantly increased in *FasL ΔIntra* mice following antigen receptor cross-linking. Surprisingly, B cell stimulation lead to a comparable extent of activating phosphorylations on S38 in c-Raf and S218/S222 in MEK1/2 in cells isolated from wildtype and *FasL ΔIntra* mice, indicating that Mitogen activated protein kinases (MAPKs) upstream of ERK1/2 (Raf-1 and MEK1/2) apparently do not contribute to the differential regulation of ERK1/2. Experiments in which activation-induced Akt phosphorylation (S473) was quantified also did not suggest a participation of Phosphoinositol specific kinase 3 (PI3K)/Akt signals in this process. Instead, further characterization of the upstream pathway revealed an involvement of Phospholipase C gamma (PLCγ) and Protein kinase C (PKC) signals in FasL-dependent ERK1/2-regulation.

Previous studies in our group revealed a Notch-like processing of FasL, resulting in the transcriptional regulation of a reporter gene. Furthermore, an interaction of the FasL ICD with the transcription factor Lymphoid-enhancer binding factor-1 (Lef-1) that affected Lef-1-dependent reporter gene transcription could be demonstrated. Therefore, a molecular analysis of activated lymphocytes was performed to identify FasL reverse signaling target genes. The differential expression of promising candidates was verified by quantitative real-time PCR (qRT-PCR), which showed that the transcription of genes associated with lymphocyte proliferation and activation was increased in *FasL ΔIntra* mice compared to wildtype mice. Interestingly, an extensive regulation of Lef-1-dependent Wnt/β-Catenin signaling-related genes was found. Lef-1 mRNA (RT-PCR) and protein (intracellular FACS staining) could be detected in mature B cells, suggesting the possibility of FasL ICD-mediated inhibition of Lef-1-dependent gene expression in these cells, initiated by Notch-like processing of FasL.

To investigate the consequences of FasL reverse signaling *in vivo*, a potential participation of the FasL ICD in the regulation of immune responses upon various challenges was analyzed. In experiments in which thymocyte proliferation or the expansion of antigen-specific T cells following a challenge with the superantigen *Staphylococcus enterotoxin B* (SEB), with Lymphocytic choriomeningitis virus (LCMV) or with *Listeria monocytogenes* were investigated, comparable results were obtained with wildtype and *FasL ΔIntra* mice. Likewise, the recruitment of neutrophils in a thioglycollate-induced model of peritonitis was not affected by deletion of the FasL ICD. These findings might reflect regulatory mechanisms operating *in vivo*, such as control exerted by regulatory T cells. Along these lines, proliferative differences in CD4$^+$ T cells could only be detected *ex vivo* after depletion of CD4$^+$CD25$^+$ regulatory T cells. Furthermore, several *in vitro* studies indicate that retrograde FasL signals can be observed under conditions of sub-optimal lymphocyte stimulation, but not when the TCR is optimally stimulated. Therefore, the potent initiation of antigen receptor signaling by stimuli like SEB or LCMV might have masked inhibitory FasL reverse signaling in these experiments.

In agreement with the observed hyperactivation of lymphocytes in the absence of the ICD *ex vivo*, the increase in germinal center B cells (GCs) following immunization with the hapten 3-hydroxy 4-nitrophenylacetyl (NP) and the number of antibody-secreting PCs was significantly higher in *FasL ΔIntra* mice. The larger quantity of PCs correlated with increased titers of NP-binding, *i.e.* antigen-specific, IgM and IgG1 antibodies in the serum of *FasL ΔIntra* mice after immunization. These data suggest that FasL reverse signaling exerts immunmodulatory functions. Supporting this notion, a model of Ovalbumin-induced allergic airway inflammation revealed an involvement of retrograde FasL-signals in the recruitment of immune effector cells into the lung and in the activation of T cells following exposure of mice to Ovalbumin.

Together, our *ex vivo* and *in vivo* findings based on endogenous FasL protein levels demonstrate that FasL ICD-mediated reverse signaling is a negative modulator of certain immune responses. It is tempting to speculate that FasL reverse signaling might be a fine-tuning mechanism to prevent autoimmune diseases, a theory which will be tested in adequate mouse models in the future.

1 Introduction

1.1 Apoptosis

The term apoptosis, often used synonymously with 'programmed cell death', describes a physiological form of cell death that proceeds through a highly ordered sequence of events and results in the self destruction of cells. In multicellular organisms, apoptosis plays crucial roles during embryonic development, in adult tissue homeostasis, the weeding out of lymphocytes after an immune response and in the killing of virus-infected cells. Paying tribute to these essential functions, its course is tightly controlled and failure in its regulation are associated with the development of severe diseases such as cancer (defective apoptosis), neurodegeneration (excess apoptosis) or autoimmune pathologies (defective clearance of apoptotic cells and/or too little apoptosis). Apoptosis is characterized by different morphological and biochemical features, including cell shrinkage, membrane blebbing, chromatin condensation and DNA fragmentation ('DNA laddering'). In a final step, so-called apoptotic bodies, loaded with cell contents, are phagocytosed by neighboring cells or macrophages to prevent an inflammatory response. This is in contrast to death by necrosis, during which leakage of cellular contents triggers a strong inflammatory reaction (Jin *et al.*, 2005; Hotchkiss *et al.*, 2009).

Apoptosis can proceed by two different mechanisms: the extrinsic and the intrinsic pathway (**Fig. 1.1**). The two pathways are triggered by different stimuli and involve a distinct apoptotic machinery. However, the cascade-like action of aspartate specific cystein proteases (caspases) is a key event in both cases, and considerable cross-talk between the extrinsic and the intrinsic pathway occurs. Caspases are synthesized as inactive pro-enzymes that become activated by (auto-) catalytic cleavage. Triggering of the apoptotic machinery leads to the auto-catalytic cleavage of initiator caspases (Caspases-2, -8, -9, -10) which subsequently convert effector caspases (Caspases-3, -6, -7) into their active form. Effector caspases then mediate the final events of cell death by targeting proteins involved in DNA repair, cell cycle, the regulation of apoptosis (anti-apoptotic proteins) or in cell architecture (Green *et al.*, 2003; Jin *et al.*, 2005; Hotchkiss *et al.*, 2009; Strasser *et al.*, 2009).

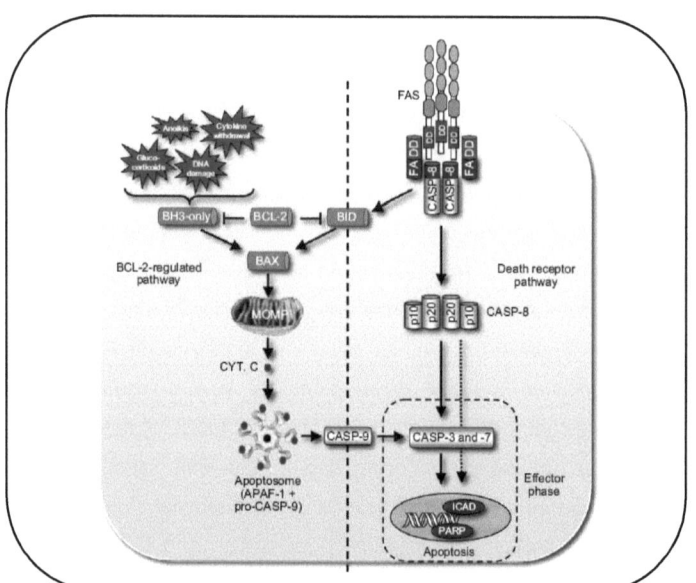

Figure 1.1 Schematic overview of extrinsic and intrinsic apoptosis pathways.
Right: The extrinsic apoptosis-pathway is triggered by death domain (DD)-containing members of the TNFR family (death-receptors; here, Fas is shown exemplary) and requires FADD-mediated activation of Caspase-8 within the DISC. Auto-proteolysis of Caspase-8 results in a heterotetrameric form that can escape the DISC and that is able to activate effector caspases (Caspase-3 and Caspase-7), eventually leading to cleavage of vital proteins within the cell. *Left:* The intrinsic Bcl-2-regulated pathway is triggered by the activation of BH3-only proteins and is balanced by the complex interplay between pro- and anti-apoptotic members of the Bcl-2 protein family. It involves mitochondrial outer membrane permeabilization (MOMP) which results in release of Cytochrome c from the intermembrane space of the mitochondria into the cytosol to initiate apoptosome formation. Within the apoptosome, Apaf-1 mediates activation of the initiator Caspase-9, further leading to proteolytic activation of the downstream effector caspases (Caspase-3 and Caspase-7). BH3-only: Bcl-2 homology domain 3-only protein; MOMP: mitochondrial outer membrane permeabilization; Cyt. C.: cytochrome c; ICAD: Inhibitor of caspase activated DNase (CAD); PARP: Poly (ADP-ribose) polymerase. Adapted from (Strasser *et al.*, 2009).

1.1.1 The extrinsic pathway of apoptosis

The extrinsic pathway of apoptosis is initiated by the binding of death ligands to their corresponding death receptors at the cell surface. Death receptor/ligand pairs belong to the tumor necrosis factor (TNF) protein family (TNFR/TNF, Fas/FasL, TRAIL-DR4 or -DR5/TRAIL). TNF-family receptors (TNFR) are type I transmembrane proteins with a characteristic ~80 aa, death domain (DD)-containing cytoplasmic tail and with motifs important for self-assembly (preligand assembly domain; PLAD) and ligand-binding (cystein-rich domains, CRDs) in the extracellular part. Notably, decoy receptors (DcR3,

DcR6) with a non-functional DD have been identified and were reported to antagonize TNFR-mediated apoptosis in a dominant negative fashion.

The cognate ligands are homotrimeric type II transmembrane proteins that bind their receptor *via* a TNF homology domain. Ligand binding induces the formation of the death-inducing signaling complex (DISC) by recruitment of DD-containing adaptor proteins like Fas-associated death domain (FADD) or TNF receptor-associated protein with a death domain (TRADD) and of pro-Caspase-8. Thereby, adaptor/receptor binding is mediated by the DD present in both proteins, and homophilic adaptor/pro-caspase interaction occurs *via* death effector domains (DED). FADD-dependent auto-catalytic processing of the pro-caspase within the DISC facilitates the Caspase-8-mediated activation of donwstream effector caspases. Caspase-8 activation at the DISC can be antagonized by cellular FLICE-like inhibtory proteins (c-FLIPs). Although common players are involved in the extrinsic pathway, their assembly and utilization is unique for each death receptor (Wajant, 2003; Eissner *et al.*, 2004; Guicciardi *et al.*, 2009; Strasser *et al.*, 2009).

1.1.2 The intrinsic pathway of apoptosis

The intrinsic pathway of apoptosis is triggered by different signals, for example DNA damage, UV radiation, growth factor withdrawl, chemicals or oxidative stress. In response to a specific death signal, distinct members of the Bcl-2-homology domain 3 (BH3)-only protein family become activated and translocate to the outer mitochondrial membrane. Their association with pro-apoptotic members of the Bcl-2 protein family results in the permeabilization of the membrane (mitochondrial outer membrane permeabilization; MOMP). Consequently, the mitochondrial membrane potential is lost, and pro-apoptotic proteins, such as Cytochrome c, inhibitor of apoptosis (IAP) antagonists and pro-caspases, are released into the cytoplasm. In the presence of dATP, the apoptosome, a ternary complex consisting of Cytochrome c, the Apoptosis protease-activating factor 1 (Apaf1) and pro-Caspase-9, is formed. Within the apoptosome, the caspase cascade is initiated by autocatalytic processing of Caspase-9 that subsequently activates effector Caspase-3. Regulation of the intrinsic pathway occurs largely through the Bcl-2 protein family: the balance between pro- and anti-apoptotic members determines the sensitivtiy of a cell to apoptosis (Shi, 2002; Jin *et al.*, 2005; Peter *et al.*, 2007; Guicciardi *et al.*, 2009; Ramaswamy *et al.*, 2009; Strasser *et al.*, 2009).

1.2 The FasL/Fas system

1.2.1 Molecular mechanism of FasL/Fas-induced apoptosis

Fas (CD95; APO-1; TNFRSF6) is a prototypic member of the TNF receptor superfamily. When expressed at the cell surface, individual Fas proteins form homotrimers *via* interaction of their PLA-domains. This pre-assembly is a pre-requisite for ligand binding (Chan *et al.*, 2000; Siegel *et al.*, 2000). Signaling *via* Fas (**Fig. 1.2**) is initiated by binding of its cognate ligand, FasL, which induces an open conformation in the cytoplasmic tails of Fas and promotes the multimerization of receptor trimers. Subsequently, DISC components are recruited to the Fas DD (Algeciras-Schimnich *et al.*, 2002; Ramaswamy *et al.*, 2009; Scott *et al.*, 2009). Cytoskeletal rearrangements and palmitoylation of Fas target the complex to glycosphingolipid-enriched membrane rafts (lipid rafts) (Chakrabandhu *et al.*, 2007; Koncz *et al.*, 2008). Upon coalescence of several lipid rafts, larger signaling platforms, so-called signaling protein oligomerization transduction structures (SPOTs), are formed to recruit pro-Caspase-8 (Siegel *et al.*, 2004). Efficient apoptotic signaling then requires receptor clustering within SPOTs ('capping') and Clathrin-mediated receptor-complex endocytosis, as endosomes serve to locally concentrate components of the apoptotic machinery and to promote DISC formation (Lee *et al.*, 2006; Chakrabandhu *et al.*, 2008; Guicciardi *et al.*, 2009).

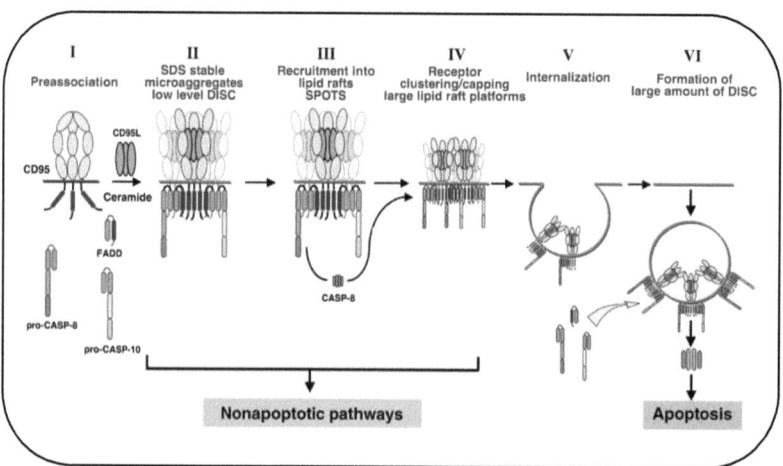

Figure 1.2 Initiation of Fas signaling.
(I) Ligand-independent receptor pre-association *via* PLADs. (II) Ligand binding triggers a conformational change, receptor multimerization and low level of DISC formation. (III) Recruitment of FasL into lipid rafts to form SPOTs. (IV) Raft coalesence. This step depends on generation of active Caspase-8 by the DISC. In the absence of internalization,

signals from steps 2-4 rather activate non-apoptotic pathways. (V) Receptor internalization by Clathrin-mediated endocytosis. (VI) Further recruitment of DISC components into endosomes and high level DISC formation. Blue domains, DED; red domain, DD; the N-terminal PLAD in Fas is shown in a different green tone. Adapted from (Lee et al., 2006).

Within the DISC, pro-Caspase-8 undergoes a FADD-mediated conformational change that facilitates its autocatalytic processing into the active, heterotetrameric enzyme. This process can be antagonized by the cellular FLICE-like inhibitory protein (c-FLIP), a caspase-like molecule without proteolytic activity that interferes with pro-Caspase-8 processing (Scaffidi et al., 1999). Mature Caspase-8 dissociates from the DISC to cleave effector Caspase-3 and/or the BH3-only protein Bid. Depending on the amount of active Caspase-8 and the actually targeted substrate, two cell types have been distinguished, type I and type II cells (Scaffidi et al., 1998; Scaffidi et al., 1999). In type I cells, Caspase-8 directly cleaves Caspase-3, and this is sufficient for effective apoptotic signaling. In contrast, type II cells require an amplification of the death receptor signal by Caspase-8-mediated cleavage of Bid. Truncated Bid (tBid) translocates to the mitochondria to trigger the caspase cascade *via* the intrinsic pathway (Andera, 2009; Ehrenschwender et al., 2009; Guicciardi et al., 2009; Ramaswamy et al., 2009; Strasser et al., 2009).

1.2.2 Apoptotic Fas signaling in the immune system

FasL/Fas-induced apoptosis is crucial for immune system homeostasis, effector functions and self tolerance and is involved in the killing of target cells by natural killer (NK) and cytotoxic T cells, in the (self) elimination of effector cells and in the maintenance of immune-privileged sites (Igney et al., 2005; Fas et al., 2006). Owing to the potent pro-apoptotic signaling capacity and the critical functions of the FasL/Fas system, FasL expression and activity are tightly regulated and restricted to few cell types, such as immune effector cells and cells of immune-privileged sites. In contrast, Fas is expressed in a variety of tissues including the lymphoid tissues, liver, heart, kidney, pancreas, brain and ovary (Kavurma et al., 2003; Guicciardi et al., 2009).

The importance of proper execution of FasL/Fas-mediated apoptosis is pointed out by two naturally occuring mutant mouse strains, $FasL^{gld/gld}$ and $Fas^{lpr/lpr}$, that harbor inactivating mutations in the *FasL* and *Fas* gene, respectively. A point mutation in the extracellular domain of FasL abolishes receptor binding in $FasL^{gld/gld}$ mice. In $Fas^{lpr/lpr}$ mice, the *Fas* gene either contains a transposon insertion in intron 2 leading to aberrant splicing, or a point mutation in the cytoplasmic DD that prohibits FADD recruitment and DISC formation

($Fas^{lpr\ cg}$). All these mutations prevent apoptosis triggered by the FasL/Fas system and lead to the production of auto-antibodies and the development of a lymphoaccumulative disease (Adachi et al., 1993; Takahashi et al., 1994a). Dominant-interfering mutations in Fas or FasL cause a similar autoimmune disease in humans called autoimmune lymphoproliferative syndrome (ALPS) (Rieux-Laucat, 2006). In mice, lymphoaccumulation involves the emergence of an anormal, 'double negative' T cell population ($CD3^+B220^+CD4^-CD8^-$) that is derived from normal $CD8^+$ T cells and that is a critical mediator of the autoimmune pathology (Cohen et al., 1991). Despite $FasL^{gld/gld}$ and $Fas^{lpr/lpr}$ mice being commonly referred to as FasL- or Fas-deficient, it could be shown that cells of both mutants retain some residual FasL and Fas activity, respectively (Adachi et al., 1993; Chu et al., 1993; Karray et al., 2004), and that conventional homozygous $FasL^{-/-}$ and $Fas^{-/-}$ knockout-mice develop a much more pronounced phenotype than the naturally occurring mutants (Karray et al., 2004; Mabrouk et al., 2008).

1.2.2.1 Lymphocyte death in the periphery: activation-induced cell death (AICD) and activated cell autonomous cell death (ACAD)

Acute infection leads to the antigen-specific activation of immune effector cells and their massive clonal expansion. These armed effector cells help to clear the infection by direct or indirect lysis of infected target cells, mediated by FasL/Fas-, perforin- or granzyme-induced cell death (Brenner et al., 2008). Once pathogens have been cleared, the immune response needs to be terminated to eliminate the now excess and potentially dangerous effector cells. While a small percentage of antigen-specific effectors becomes memory cells, retained as safeguards against future infections, most cells are deleted by suicide or fratricide.

Once thought to be dependent on Fas, cell death following acute immune responses is attributed to activated cell autonomous death (ACAD) now. This death by neglect depends on the BH3-only protein BIM and occurs due to the relative cytokine deficiency after clonal expansion that leads to a shift in the balance of pro- and anti-apoptotic proteins (Hildeman et al., 2002; Green et al., 2003; Fas et al., 2006; Krammer et al., 2007; Brenner et al., 2008).

In contrast, activation-induced cell death (AICD), also sometimes referred to as re-stimulation-induced cell death (RICD), is responsible for the elimination of cells in states of antigen persistance, for example in chronic infections. Persistent antigen re-stimulates already activated lymphocytes and thus, renders lymphocytes sensitive to apoptosis. AICD often, but not necessarily, is induced by death receptor signaling, and the concomitant increases of FasL cell surface expression upon re-stimulation and of sensitivity to Fas-mediated killing are crucial mediators (Stranges *et al.*, 2007; Ramaswamy *et al.*, 2009). AICD not only plays a role in chronic infections, but has also been implicated in the maintenance of peripheral tolerance. Self-reactive lymphocytes are repeatedly stimulated by the frequent encounter of (self) antigen, rendering them sensitive to AICD, especially in the absence of co-stimulation (Brenner *et al.*, 2008; Strasser *et al.*, 2009).

1.2.2.2 FasL/Fas-induced apoptosis in B cell function
Besides its role in AICD, Fas-induced apoptosis of or by B cells is crucial for proper immune system functions and the prevention of autoimmunity. Antigen-activated B cells upregulate Fas, but are initially protected from apoptosis due to simultaneous cross-linking of their B cell receptor (BCR) and of the co-stimulatory receptor CD40. Upon immune response termination, T cells downregulate CD40 ligand (CD40L) expression and display increased levels of FasL. Consequently, B cells lose the co-stimulatory signal, become Fas-sensitive and are deleted by FasL-expressing T cells (Rathmell *et al.*, 1996; Mizuno *et al.*, 2003; Ramaswamy *et al.*, 2009). Likewise, B cell selection in germinal centers depends on apoptotic Fas signaling. Within germinal centers, B cells undergo somatic hypermutation, receptor editing and immunglobulin (Ig) isotype class switching, resulting in high affinity antibody-producing plasma cells. Providing an important mechanism for maintenance and induction of peripheral tolerance, B cells that become auto-reactive during these mutational events or fail to produce high affinity antibodies are eliminated by FasL-expressing intrafollicular CD4[+] T cells (Rathmell *et al.*, 1996; Guzman-Rojas *et al.*, 2002; McHeyzer-Williams, 2003; Mizuno *et al.*, 2003).

In addition to receiving an apoptotic Fas signal, B cells themselves can deliver such a signal *via* FasL. Induction of FasL expression on B cells is observed upon bacterial, parasitic or viral infections, such as infection with *Schistosoma mansoni*, *Trypanosoma cruzi*, human immunodeficiency virus (HIV), murine leukemia virus (MuLV) or Epstein-Barr virus (EBV). Furthermore, FasL-expressing B cells control diseases caused by de-regulated T cells, and this is involved in peripheral tolerance and autoimmunity, for example during airway inflammation-, asthma- or transplant-induced immune responses. In most instances, B cells were shown to eliminate activated, antigen-specific T cells *via* FasL-mediated AICD. This process might be aided by the capacity of B cells to serve as professional antigen presenting cells (APCs) for T cells (Hahne et al., 1996; Lundy et al., 2002; Bonardelle et al., 2005; Lundy, 2009; Lundy et al., 2009).

Interestingly, fully matured plasma cells, i.e. those having undergone class switching to IgA or IgG and secreting quantities of antibody, were shown to store high amounts of FasL in para-nuclear, probably Golgi-associated compartments. It was hypothesized (without direct experimental evidence) that plasma cells delivering antibodies to the site of infection concomitantly express FasL at their cell surface, or secrete exosome-associated FasL, to delete local effector T cells and to down-modulate the immune response (Strater et al., 1999).

1.2.3 Apoptotic Fas signaling outside the immune system

FasL/Fas signaling contributes to tissue homeostasis and pathologic states in some non-immune tissues. Fas-induced apoptosis was shown to play a role in liver homeostasis, as Fas-deficient mice develop liver hyperplasia. In the neurological system, FasL/Fas-mediated apoptosis of neurons occurs as a consequence of growth factor withdrawl and spinal cord injury, is induced by the Alzheimer's disease associate Aβ-peptide and plays a key role in neonatal brain disorders. Furthermore, pancreatic islet cell destruction in type I diabetes depends on Fas, and in the cardiovascular system apoptotic Fas signaling has been implicated in the regulation of cardiac hypertrophy, angiogenesis and the response to reperfusion injury in stroke and myocardial infarction (Green et al., 2003; Ehrenschwender et al., 2009; Hotchkiss et al., 2009).

1.2.4 FasL/Fas signaling in the establishment of immune privilege and in tumor biology

Due to its very potent pro-apoptotic potential, FasL expression is restricted to immune effector cells and a few tissues outside the immune system, including the eye, brain, testis and uterus (French et al., 1996a; French et al., 1996b; Nagata, 1997; Kavurma et al., 2003). The non-lymphoid FasL-expressing tissues are special as they do not reject allogenic transplants and, therefore, are said to be 'immune-privileged'. The immune-privileged status largely depends on the capacity to eliminate infiltrating lymphocytes by FasL/Fas-induced apoptosis to prevent tissue damage and a local inflammatory reaction (Griffith et al., 1995; Itoh et al., 1995; Streilein, 1995). However, the capacity to kill in a FasL-dependent manner is not the sole determinant of immune privilege, and production of immunsuppressive cytokines, low levels of major histocompatibility (MHC) complex expression and a physical barrier are required concomitantly (Green et al., 2001).

Besides the constitutive immune-privileged status of some tissues, induced immune privilege has been discussed for tissues like the small intestine, lung, liver and skin, as a mechanism to attenuate or terminate local immune responses and to prevent excess tissue damage (Schwarz et al., 1998; Green et al., 2001; Viard-Leveugle et al., 2003). Induced immune privilege involves the induction of FasL expression at the surface of non-lymphoid cells to eliminate infiltrating T cells by AICD. It is hypothesized, that FasL expression is mediated by the T cells themselves: activated T cells produce TNFα, which is known to induce FasL. In this scenario, an immune-privileged status would only be obtained above a certain threshold of T cell signaling strength; below that threshold, peripheral T cells would not be deleted, and the immune response could proceed unhindered (Green et al., 2001; Krammer et al., 2007).

Using similar mechanisms as the ones used to establish natural immune-privileged sites, tumors can establish a quasi immune-privileged status for themselves. Many tumor cells express FasL and/or secrete FasL-containing microvesicles and were found to kill infiltrating, activated lymphocytes (Igney et al., 2005). This concept of a 'tumor counterattack' is controversial and has been challenged by the observation of FasL$^+$ transplant rejection, the higher incidence of plasmacytoid lymphomas in Fas- and FasL-defective mice ($Fas^{lpr/lpr}$, $FasL^{gld/gld}$) and of enhanced inflammation upon Fas stimulation at the surface of tumor cells (Bellgrau et al., 1995; Davidson et al., 1998; Brunner et al., 2001; Wasem et al., 2001; Allison et al., 2005; Ehrenschwender et al., 2009). However, inflammatory neutrophil recruitment has also been associated with tumor growth, possibly due to non-apoptotic Fas signaling (see below) mediated by FasL expressed on neutrophils (Restifo, 2000). Indeed, many tumors are resistant to Fas-mediated apoptosis, and non-apoptotic Fas signaling triggered by neighboring tumor cells or by infiltrating immune cells has been correlated with tumor motility and invasiveness in vitro (Barnhart et al., 2004; Ametller et al., 2010). A recent report demonstrates in various human cancer cell lines in vitro and in mouse models of ovarian and liver cancer in vivo that growth and survival of many cancer cells depend on constitutive, non-apoptotic Fas signaling (Chen et al., 2010). Intriguingly, these signals are delivered by tumor-produced FasL, not by FasL-expressing cells in the microenvironment or of the immune system, and require only very low levels of Fas and FasL protein expression.

Thus, there might be a dual role for FasL/Fas in cancer: in some instances it might aid in tumor surveillance and regression, while in others, non-apoptotic Fas-signaling into tumors and/or FasL-mediated killing of immune effectors by tumor cells ('counterattack') favour tumor growth and aggressiveness (Green et al., 2001; Igney et al., 2005; Peter et al., 2007).

1.2.5 Non-apoptotic Fas signaling

In addition to its potent pro-apoptotic function, the FasL/Fas system can elicit non-apoptotic signals in the receptor-bearing cell as well. Fas signaling was shown to exert co-stimulatory functions, such as promotion of proliferation, cytokine production and survival of mature lymphocytes and thymocytes, of positive selection in the thymus and of dendritic cell maturation (Wajant et al., 2003; Guicciardi et al., 2009). Additionally, proliferation of serum-starved fibroblasts depends on Fas, and liver regeneration after partial liver hepatectomy is blocked in Fas-deficient models. In the neural system, non-apoptotic Fas signaling induces neurite outgrowth and branching, mediates nerve regeneration and contributes to adult neuronal stem cell maintenance and function. Further known functions of non-apoptotic Fas signaling include contributions to the homeostasis of intestine, heart and pancreas and to oncogenic signaling leading to tumor progression (Janssen et al., 2003; Wajant et al., 2003; Barnhart et al., 2004; Ehrenschwender et al., 2009; Strasser et al., 2009; Chen et al., 2010).

Which signals determine the consequences (pro- vs. anti-apoptotic) of Fas signaling? Although the precise mechanism of decision making has yet to be elucidated, it is clear that FasL/Fas signaling requires more than the mere ligand/receptor interaction and that the subcellular localization, post-translational modifications, strength of receptor triggering and the microenvironment, including the activity levels of intracellular signaling pathways, all play crucial roles. Along these lines, post-translational modifications of Y291 and/or C199 in the Fas intracellular domain control Fas multimerization, targeting to lipid rafts, receptor internalization and the route of receptor internalization (Clathrin-dependent vs. -independent). Retention of Fas at the cell surface and Clathrin-independent internalization both favour non-apoptotic outcomes (Lee et al., 2006; Chakrabandhu et al., 2007; Chakrabandhu et al., 2008; Rossin et al., 2010).

c-FLIP is discussed as another important player in the regulation of Fas signals. Well described for its blockade of apoptosis by inhibiting Caspase-8 processing, it has also been shown to exert pro-apoptotic functions by activating Caspase-8 and, on the other hand, to trigger pro-survival and pro-proliferative pathways like the Nuclear Factor-κb (NF-κb)-, the Mitogen-activated protein kinase (MAPK) and the Phosphoinositide 3-kinase

(PI3K)/Akt-pathway (Hu et al., 2000; Dohrman et al., 2005; Yu et al., 2008a; Strasser et al., 2009): MAPK signaling is probably initiated by binding of c-FLIP to Raf-1; NF-κb activation requires pro-Caspase-8-mediated cleavage of FLIP into p43-FLIP and p22-FLIP, leading to direct or indirect degradation of the I-κb kinase (IKK) complex. Recently, a direct interaction between c-FLIP and Akt has been shown to promote anti-apoptotic Akt functions (Krammer et al., 2007; Brenner et al., 2008; Guicciardi et al., 2009; Quintavalle et al., 2010).

In addition, the DISC components FADD and Caspase-8 link Fas to anti-apoptotic pathways, e.g. to the MAPK/Extracellular signal regulated kinase (ERK1/2)-, NFκB-, PI3K-, Jun N-terminal kinase (JNK) and p38 pathway. The exact roles of FADD and Caspase-8 are not clear, and it is debated whether enzymatic caspase activity is required. It seems likely that the strength of receptor triggering determines the level of Caspase-8 activation and localization. While mature Caspase-8 heterotetramers escape the DISC to initiate the caspase cascade, un- or not fully processed Caspase-8 has been shown to remain localized to the DISC and to be involved in non-apoptotic signaling. Additionally, FADD and Caspase-8 have been found in a complex with the kinase Receptor interacting protein 1 (RIP1). Post-translational modifications of RIP-1 are decisive for survival vs. death signaling (Bentele et al., 2004; Krammer et al., 2007; Lavrik et al., 2007; Matsumoto et al., 2007; Peter et al., 2007; Guicciardi et al., 2009; Strasser et al., 2009; Chen et al., 2010).

Interestingly, the Fas DD contains a motif (YXXL) that resembles immunoreceptor tyrosine activation motifs (ITAMs) or immunoreceptor tyrosine inhibitory motifs (ITIMs) and that allows the binding of SH2 domain-containing proteins. Tyrosine phosphorylation within this motif correlates with activation of the Src-family kinases (SFK) Lyn and Fyn and with activation of the PI3K pathway. This tyrosine phosphorylation facilitates recruitment of the SFK-targeting Protein tyrosine phosphatase 1/2 (SHP1/2) and SH2-domain containing inositol polyphosphate 5'-phosphatase (SHIP) in T cells and glioma cells. Apparently, the outcome of Fas signaling is also balanced by a kinase-phosphatase interplay (Corsini et al., 2009; Sancho-Martinez et al., 2009).

1.3 The Fas Ligand

1.3.1 FasL structure

Fas Ligand (FasL; CD95L; CD178; TNSF6) is a 40 kDa glycosylated type II transmembrane protein with 279 aa in mice and 281 aa in humans that belongs to the TNF family (**Fig. 1.3**). It is conserved across species with a sequence identity of 76.9% in mice vs. humans (Suda et al., 1993; Takahashi et al., 1994b). The extracellular domain (ECD) harbors a TNF homology domain, the receptor binding site, a motif for self assembly and trimerization and several putative N-glycosylation and a metalloprotease cleavage site/s (Orlinick et al., 1997a; Orlinick et al., 1997b). FasL shedding by metalloproteases liberates soluble FasL (sFasL). Alternatively, sFasL is produced by splicing. The cytoplasmic tail of FasL is the longest of all TNFL family members and contains several conserved signaling motifs, such as a putative tandem Casein kinase I phosyphorylation site (CKI-S), a unique proline-rich domain (PRD) and, in mice, one potential tyrosine phosphorylation site at aa 7. Human FasL can be phosphorylated at three tyrosine residues: aa 7, 9, and 13 (Hahne et al., 1995; Watts et al., 1999; Voss et al., 2008; Ehrenschwender et al., 2009; Weinlich et al., 2010).

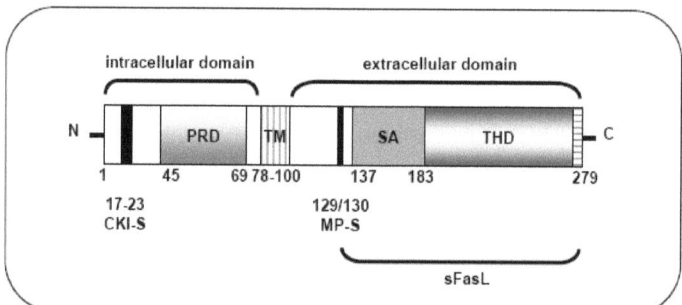

Figure 1.3 Structure of the murine FasL.
The C-terminal extracellular domain contains the receptor binding site (horizontal stripes), a region for self assembly (SA) and the TNF homology domain (THD) which contains potential N-glycosylation sites. Proteolysis at the metalloprotease cleavage site (MP-S) liberates soluble FasL (sFasL). The N-terminal intracellular domain harbors a unique proline-rich domain (PRD) and a tandem Casein kinase I phosphorylation site (CKI-S). Murine FasL has one tryosine residue that can be phosphorylated (Y7), while human FasL contains three phosphorylatable tyrosines (Y7, Y9, Y13). Modified from (Voss et al., 2008).

1.3.2 Regulation of FasL expression and activity

Unlike Fas, FasL is only found in few cell types where its expression and activity is tightly regulated at the transcriptional level and by several post-translational modifications. While FasL is constitutively expressed in non-lymphoid, immune-privileged tissues, it is present at very low levels or completely absent at the surface of resting lymphocytes. FasL only becomes upregulated upon antigen-receptor stimulation in T and B cells (French et al., 1996a).

1.3.2.1 Transcriptional control

Several cis-acting promoter elements have been implicated in the positive or negative regulation of the FasL gene. Activation-induced transcription is mediated by classical transcription factors like Stimulating protein-1 (SP-1), Nuclear factor of activated T cells (NFAT), NF-κb, Activating protein 1 (AP-1), Early growth response-3 (Egr-3), c-Myc, Interferon regulatory factor 1 (IRF-1) and the Cyclin B1/Cyclin-dependent kinase 1 (Cdk 1) complex. Negative regulators of FasL expression that antagonize aforementioned transcriptional activators include MHC class II transactivator (CIITA), retinoic acid, Transforming growth factor-β (TGF-β) and nitric oxide (Lettau et al., 2009).

1.3.2.2 FasL sorting and storage

The localization of FasL is presumed to differ between hematopoietic and non-hematopoietic cells. In hematopoietic cells FasL is stored intracellularly in multivesicular bodies and becomes externalized upon cell activation only, while non-hematopoietic cells constitutively express FasL at the cell surface (French et al., 1996a; Kavurma et al., 2003). According to a publication by Griffith and co-workers, FasL is sorted to so-called secretory lysosomes after its synthesis (Bossi et al., 1999). In the human system this was shown to require Src-family kinase-mediated lysine mono-ubiquitination and phosphorylation of tyrosine residues in the FasL intracellular domain (ICD) (Zuccato et al., 2007). It is further assumed that lysosomal sorting critically depends on the FasL PRD, since deletion of the cytoplasmic tail leads to the constitutive expression of FasL on the cell surface of hematopoietic cells (Blott et al., 2001). However, recent studies did not find a co-localization of FasL with classical markers (CD63, CD107a) or components (perforin, granzyme) of secretory lysosomes, suggesting FasL storage in other compartments (He et al., 2007; Kassahn et al., 2009; Schmidt et al., 2009).

Lymphocyte stimulation *via* the antigen receptor triggers a biphasic expression of FasL at the cell surface. Shortly after activation, preformed FasL is released at the immunological synapse and aids in the locally restricted, highly specific killing of target cells (Lettau, 2004). Additionally, lysosomal degranulation releases microvesicular bodies in which FasL is embedded in the plasma membrane (Martinez-Lorenzo *et al.*, 1999). The second wave of expression requires *de novo* synthesis and might involve direct transport of FasL to the cell surface, as it was shown to be independent of lysosomal degranulation. FasL de *novo* synthesis is controlled in at least two ways: Xiao and colleagues reported a negative regulation of FasL synthesis by aa 2-33 in the FasL cytoplasmic tail (Xiao *et al.*, 2004). Recently, it was demonstrated that induction of *de novo* synthesis depends on strong antigen receptor-mediated signals and that release of newly synthesized FasL is associated with significant levels of bystander cell lysis. In contrast, weak receptor-signals trigger the delivery of stored FasL specifically to the immunological synapse and do not induce *de novo* synthesis (He *et al.*, 2010).

1.3.2.3 Regulation of FasL activity by dynamic localization in lipid rafts

As described for the Fas receptor (Hueber *et al.*, 2002), a fraction of FasL is constitutively localized in membrane lipid rafts (Gajate *et al.*, 2005; Cahuzac *et al.*, 2006). In a collaboration with Dr. Anne-Odile Hueber's laboratory (INSERM/CRNS, Nice, France), the group of PD Dr. Martin Zörnig could show that increased amounts of FasL are found in rafts after efficient FasL/Fas interaction (Cahuzac *et al.*, 2006). Raft disorganization and deletions within the intracellular FasL domain, particularly removal of the PRD, diminish raft partitioning and lead to decreased FasL killing (Cahuzac *et al.*, 2006; Nachbur *et al.*, 2006). Thus, FasL raft partitioning seems to be essential for maximum contact with the Fas receptor and the ability of FasL to induce Fas-mediated apoptosis.

It has been suggested that targeting of FasL to rafts depends on palmitoylation of a cystein residue in the very N-terminal region of the transmembrane domain (aa 82). In this study, a C82S mutation abolishes raft partitioning, FasL cleavage by ADAM10 and target cell killing (Guardiola *et al.*, manuscript in preparation). However, further analysis of the physiological relevance is needed, as the palmitoylatable cystein residue is not conserved. While it is present in human and porcine FasL, murine and rat proteins harbor tryptophan at position 82 instead.

1.3.2.4 FasL processing

Upon expression at the cell surface, the extracellular domain of FasL can be shed by matrix metalloprotease activity to produce sFasL. The function of sFasL was discussed controversial, and sFasL was reported to either cause apoptosis at a distant site, to dampen the apoptotic response by blocking Fas receptors or to be inactive (Tanaka et al., 1995; Suda et al., 1997; Schneider et al., 1998; Tanaka et al., 1998; Knox et al., 2003). A recent study using knockout/knockin mouse models largely solved this issue by demonstrating that only membrane-bound FasL mediates target cell lysis and that sFasL rather provokes pro-inflammatory and tumorigenic effects (O' Reilly et al., 2009).

Recently, we have described a Notch-like proteolytic processing of FasL (**Fig. 1.4**). In a first step, membrane FasL is cleaved by the metalloprotease ADAM10 to generate sFasL and a 17 kDa fragment which remains membrane-anchored (Kirkin et al., 2007; Schulte et al., 2007). As discussed above, this step might require FasL palmitoylation and raft partitioning (Guardiola et al., manuscript in preparation). A subsequent cleavage within the transmembrane region by the protease Signal peptide peptidase-like 2a (SPPL2a) releases the intracellular domain (ICD) into the cytoplasm. The FasL ICD then translocates to the nucleus and may influence gene transcription (Kirkin et al., 2007).

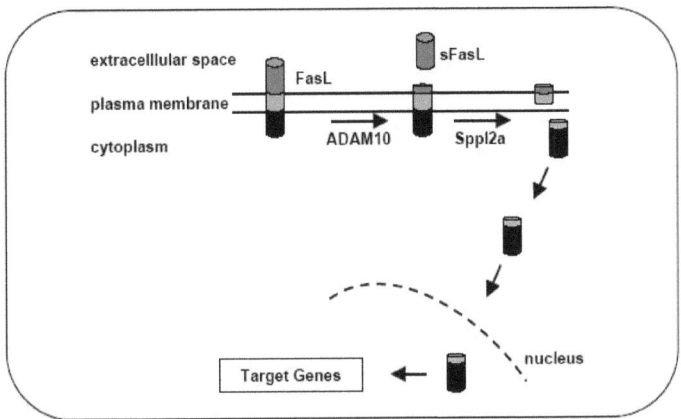

Figure 1.4 Proteolytic FasL processing.
The FasL ECD is shed by the protease ADAM10 to generate sFasL and a membrane-anchored fragment consisting of TM and ICD. A second cleavage by the intramembrane-protease SPPL2a liberates the ICD which is able to translocate to the nucleus to regulate target gene expression. Black: intracellular domain (ICD); light grey: transmembrane domain (TM); dark grey: extracellular domain (ECD)

1.3.2.5 Regulation of FasL by proteins that bind to its intracellular domain

Several FasL binding partners have been identified by GST-pull downs, phage display technology or in Yeast-two-hybrid screens, and the interaction with FasL was verified for a number of these proteins by co-immunoprecipitation experiments. The majority of the FasL binding proteins interact with the FasL PRD domain *via* their SH3 or WW domains and regulate many aspects of FasL biology (Baum *et al.*, 2005; Qian *et al.*, 2006; Thornhill *et al.*, 2007; Thornhill *et al.*, 2008; Voss *et al.*, 2009). Putative and verified interaction partners include Src-family kinases (Src, Fyn, Lyn, Lck, Hck, Fgr, Abl), adaptor proteins (Grb2, Gads, p85 subunit of PI3K, Nck) and pombe cdc15 homology (PCH) family members (e.g. CD2BP1/PSTPIP, PACSIN1-3, FBP17, CIP4), and they link FasL to various intracellular signaling pathways. Src-kinases and adaptors like Growth factor binding protein 2 (Grb2) and Grb2-related adaptor downstream of Shc (Gads) are important mediators of antigen receptor-stimulated lymphocyte activation and could fascilitate FasL reverse signaling by connecting FasL and antigen receptor signals. Additionally, Src-kinases mediate the phosphorylation of FasL required for proper lysosomal sorting (Zuccato *et al.*, 2007). The adaptor protein Nck is involved in activation-induced reorganization of the actin cytoskeleton in lymphocytes, and binding of Nck to the PRD was shown to facilitate recruitment of FasL-containing vesicles to the immunological synapse for targeted delivery (Lettau *et al.*, 2006). PCH-family members have been implicated in the regulation of FasL localization: The group of PD Dr. Martin Zörnig reported a ternary complex consisting of FasL, the SH3 domain-containing adaptor protein Proline, serine, threonine phosphatase interacting protein (PSTPIP) and the Protein tyrosine phosphatase-PEST, non-receptor type, 12 (PTP-PEST; PTPN12). According to the data, formation of this complex reduces FasL cell surface expression by promoting intracellular storage of the ligand which correlates with a decreased FasL-dependent killing capacity (Baum *et al.*, 2005). Furthermore, this complex is an attractive candidate for the initiation of FasL reverse signaling, since PSTPIP and PTP-PEST have been shown to negatively affect lymphocyte activation (Badour *et al.*, 2003; Yang *et al.*, 2006).

Adding another facette to this already complex picture, we could recently confirm an interaction of the FasL ICD with the transcription factor Lymphoid-enhancer binding factor-1 (Lef-1) that regulates Lef-1-dependent transcription *in vitro* (Lückerath *et al.*, manuscript submitted).

1.4 Reverse ligand signaling

1.4.1 Reverse signal-transduction by TNF family ligands

Up to now, 19 TNF family ligands have been identified, most of which are synthesized as type II transmembrane proteins and play important roles in immune system functions (Bodmer *et al.*, 2002). Binding to cognate TNF-receptors is mediated by their extracellular TNF homology domain. Although the existence of soluble forms has been demonstrated for most TNFLs, receptor binding and triggering predominantely depends on the membrane-anchored ligand (Aggarwal, 2003; Matthies *et al.*, 2006). In addition to the well-described capacity to deliver a signal into receptor-bearing cells, many TNF ligands can receive signals themselves, a phenomenon which is known as 'reverse signal-transduction'. Among the ligands for which a reverse signaling has been reported are CD40L (CD154), mTNF, FasL, OX40L, CD70 (CD27L), 4-1BB-L (CD137L), TNF-related activation-induced cytokine receptor (TRANCE; RANKL), TNF-related apoptosis-inducing ligand (TRAIL), CD30L (CD153) and LIGHT (**Table 1.1**). Most studies found a (co-) stimulatory potential during antigen receptor-induced lymphocyte activation, proliferation and cytokine production. However, inhibitory functions have been described as well, and the outcome of retrograde signal-transduction probably depends on the cellular context. Along these lines, the precise molecular mechanism involved is not clear. As the cytoplasmic tail of TNFLs is conserved across species, but not between different family members, and since it harbors signaling motifs, e.g. a CK-I substrate site in CD40L, mTNF, FasL, CD70, 4-1BB-L and CD30L, retrograde signal-transduction is likely mediated by the respective intracellular domain (Watts *et al.*, 1999; Eissner *et al.*, 2004; Sun *et al.*, 2007a).

Of note, bidirectional signaling is not restricted to TNF family members but has also been described for Notch (D'Souza *et al.*, 2008), ephrins (Pasquale, 2008), EpCam (Maetzel *et al.*, 2009), Interleukin-15 (Bulfone-Paus *et al.*, 2006), CD47 (Matozaki *et al.*, 2009), semaphorins (Zhou *et al.*, 2008) and possibly others. Intriguingly, there is a striking similarity in the mechanisms underlying reverse signaling and its regulation.

Table 1.1 TNFL members for which reverse signaling has been described.
BC : B cell ; DC : dendritic cell ; NK: NK cells; TC: T cell ; MΦ: macrophages; M: monocytes; PMN: polymorphonuclear granulocyte (neutrophils); modified after (Sun et al., 2007a).

Name	Expression	Receptor	Implications of reverse signaling
CD40L	activated TC	CD40	co-stimulation, direct stimulation
mTNF	MΦ, M	TNFR2	co-stimulation, desensitisation
FasL	activated TC & NK, BC	Fas; DcR3	co-stimulation, co-inhibition
OX40L	activated TC, BC, DC, M	OX40	direct stimulation
CD70	activated TC, BC, DC, M	CD27	direct stimulation
4-1BB-L	activated TC, BC, DC, activated M	4-1BB	co-stimulation of B cells ; activation, migration and infiltration of monocytes; inhibition of osteoclastogenesis and T cell proliferation (Senthilkumar et al., 2009)
TRANCE	activated TC	RANK	co-stimulation; inhibition of osteoclastogenesis and T cell proliferation (Senthilkumar et al., 2009)
TRAIL	activated TC & NK	TRAILR1	co-stimulation
CD30L	PMN, BC, MΦ, activated TC	CD30	co-stimulation, direct stimulation, inhibition
LIGHT	activated TC, immature DC, M	HVEM, LTαβR, DcR3	co-stimulation

1.4.2 FasL reverse signaling

The first evidence for the existence of reverse signaling into the FasL-bearing cell stems from two studies that demonstrated either co-stimulation of murine CD8$^+$ T cell lines by FasL cross-linking (Suzuki et al., 1998) or inhibition of activation-induced proliferation of murine CD4$^+$ T cell lines (Desbarats et al., 1998). In both cases, T cell lines derived from *FasL$^{gld/gld}$*, *Fas$^{lpr/lpr}$* and wildtype mice were analyzed, and it could be shown that FasL-deficient *FasL$^{gld/gld}$* cells responded differentially to concomitant FasL and TCR engagement compared to wildtype and *Fas$^{lpr/lpr}$* cells. These findings revealed that co-stimulation or inhibition by FasL critically depended on the presence of signaling-competent ligand, and they support the existence of a true reverse signaling into the ligand-bearing cell (instead of FasL-mediated receptor signals ('forward signals')).

Almost certainly, the FasL ICD is functionally involved in transmission of this signal: (i) The ICD is highly conserved across species and harbors several signaling motifs, such as a unique PRD, a potential Casein kinase I substrate site, and it contains phosphorylatable tyrosines (Y7 in mouse; Y7, Y9, Y13 in human). (ii) Numerous proteins that bind to the PRD have been identified. These interactions aid in the regulation of FasL sorting, storage and cell surface expression, and they link FasL to intracellular signaling pathways (Lettau et al., 2009; Ramaswamy et al., 2009; Weinlich et al., 2010). (iii) Post-translational modifications of the ICD have been implicated in the sorting of FasL to vesicles (Zuccato et al., 2007), and targeted mutations of serines that can be phosphoylated by CK-I decreased the activity of a Luciferase reporter construct for NFAT activation (Sun et al., 2006). (iv) Proteolytic processing of FasL liberates the ICD and allows its translocation into the nucleus where it might influence gene transcription (Kirkin et al., 2007). (v) Sun and co-workers (Sun et al., 2006) demonstrated that expression and cross-linking of a membrane-tethered FasL ICD construct sufficed to initiate reverse signaling.

Interestingly, bi-directional FasL signaling has been observed in non-lymphoid cells as well. In a murine sertoli cell line, FasL engagement proved to be essential for MAPK/ERK1/2-dependent activation of cytosolic Phospholipase A2 ($cPLA_2$) (Ulisse et al., 2000).

1.4.3 Functional implications of FasL reverse signaling

Conflicting data about the consequences of FasL reverse signaling exist, and co-stimulatory as well as inhibitory functions have been reported. Employing $CD8^+$ T cell lines derived from $FasL^{gld/gld}$ mice, one group found an impaired proliferative capacity of these cells upon T cell receptor (TCR) cross-linking, compared to wildtype or $Fas^{lpr/lpr}$ $CD8^+$ T cells (Suzuki et al., 1998). In contrast, TCR-stimulation-induced proliferation of wildtype and $Fas^{lpr/lpr}$ $CD4^+$ T cells, but not of $FasL^{gld/gld}$ $CD4^+$ T cells, could be inhibited by immobilized Fas-Fc fusion proteins, and this was accompanied by inhibition of IL-2 production and cell cycle progression (Desbarats et al., 1998).

Follow-up studies by the group of Pamela Fink collected data supporting FasL-mediated co-stimulation, for example during positive selection of thymocytes (Boursalian et al., 2003) and the proliferative response of mature $CD8^+$ T cells (Suzuki et al., 2000b; Sun et al., 2006; Sun et al., 2007b). Contradicting the publication by Desbarats et al. (1998), they claimed that both, naïve $CD4^+$ T cells and $CD8^+$ T cells, receive co-stimulatory FasL signals. In their hands, AICD masks the co-stimulatory signal in $CD4^+$ T cells in later phases of the immune response, when CD4$^+$ T cells become Fas sensitive while still expressing FasL. This does not occur in CD8$^+$ T cells, as CD8$^+$ T cells downregulate FasL and are comparably resistant to Fas-mediated death (Suzuki et al., 2000a).

According to their findings, reverse signaling requires concomitant engagement of FasL and the TCR. This recruits FasL into lipid rafts, where it associates with SH3 domain-containing proteins, such as Fyn, Grb2 and the p85 subunit of PI3K. The Akt pathway is then triggered via p85-PI3K leading to ERK1/2 activation, proliferation and IFN-γ production. Similarly, the transcription factor AP-1 is activated via the JNK pathway (Sun et al., 2006). It could be shown, that co-stimulation necessarily depends on the PRD, precisely on the SH3 domain binding site (aa 45–54) within the PRD. Notably, activation and nuclear translocation of NFAT apparently do not rely on the presence of the PRD, but require serine phosphorylation by CK-I (Sun et al., 2007b).

However, the notion of FasL-mediated co-stimulation has again been challenged by a recent publication, in which FasL cross-linking by plate-bound Fas-Fc inhibits TCR-triggered activation and proliferation of CD4$^+$ and CD8$^+$ T cells isolated from peripheral human blood (Paulsen et al., 2009). This report suggests that FasL reverse signaling regulates TCR-induced signal-transduction at the very early step of TCR internalization. Consequently, activation of downstream signaling molecules, such as ζ-chain-associated protein kinase of 70 kDa (ZAP-70), Linker for activation of T cells (LAT), Phospholipase C gamma 1 (PLCγ1) and ERK1/2, is blocked, thus attenuating T cell activation upon antigen-encounter.

1.5 Aims of the project

The conflicting data about the consequences of FasL reverse signaling probably reflect the use of artificial experimental systems. Neither the precise molecular mechanism underlying FasL reverse signaling, nor its physiological relevance have been addressed at the endogenous protein level *in vivo*. Therefore, a 'knockout/knockin' mouse model in which wildtype FasL was replaced with a deletion mutant lacking the intracellular portion (*FasL ΔIntra*) was established in the group of PD Dr. Martin Zörnig by Dr. Vladimir Kirkin. In the present study, *FasL ΔIntra* mice are phenotypically characterized and are employed to investigate the physiological consequences of FasL reverse signaling on the molecular and cellular level.

Issues addressed include:

(i) Do *FasL ΔIntra* mice indeed represent a suitable model to study the consequences of FasL reverse signaling?
FasL expression and functionality will be compared in T and B cells derived from *FasL ΔIntra* and wildtype mice. Systematic phenotypic screening performed at the German Mouse Clinic (GMC, Munich) and in our laboratory should reveal any phenotypic abnormalities that might develop due to defective reverse signal-transduction.

(ii) Does FasL reverse signaling play a role in the regulation of lymphocyte proliferation *ex vivo* and *in vivo*?
Proliferation of lymphocytes in response to antigen receptor stimulation will be analyzed. If retrograde FasL signals participate in the regulation of proliferation, the underlying molecular mechanism will be elucidated.

(iii) Which target genes are influenced by FasL reverse signaling?
Molecular phenotyping of activated lymphocytes will be performed by *Affymetrix* gene profiling. The differential expression of potential target genes will be verified by quantitative real time PCR (qRT-PCR).

(iv) What are the consequences of FasL reverse signaling *in vivo*?
A potential participation of FasL reverse signaling in the regulation of immune responses to various challenges will be investigated.

2 Materials and methods

2.1 Materials

2.1.1 Equipment

Autoclave Tuttnauer Systec 2540 EL	Systec, Wettenberg
BioRad Power PAC 300	BioRad, Munich
Branson Sonifier 250	Heinemann, Schwäbisch Gmünd
CASY® Cell Counter	Schärfe System, Reutlingen
Centrifuge Megafuge 1.0R	Heraeus, Hanau
Centrifuge Micro 220R	Hettich Zentrifugen, Tuttlingen
Centrifuge Minifuge GL	Heraeus, Hanau
DNA sub cell™, for agarose gels	BioRad, München
Electrophoresis power supply EPS 301	GE healthcare, Freiburg
FACSCalibur	Becton Dickinson, Heidelberg
FluoStar Optima ELISA reader	BMG Labtechnologies, Offenburg
Freezer CFC free (-80°C)	Sanyo, Wiesloch
Fridge (4°C) and freezer (-20°C)	Liebherr Int., Ochsenhausen
SemiPhor, Semi dry blotting device	Pharmacia, Freiburg
Hypercassette™	GE healthcare, Freiburg
Incubator IR-sensor (37°C, 5% CO_2)	Sanyo, Wiesbaden
Laminar air flow cabinet	Clean Air, Woerden
Light Cycler 480	Roche, Mannheim
MACS magnet	Miltenyi, Bergisch Gladbach
Magnetic stirrer IKA RCT basic	IKA Labortechnik, Staufen
Micowave oven	Samsung, Berlin
Microscope (cell culture)	Hund, Wetzlar
Mini-Protean® 3 Cell gel chamber	BioRad, München
Nanodrop 1000 spectrophotometer	Thermo Scientific, Bonn
Optimax Typ TR film developer	MS Laborgeräte, Wiesloch
pH meter FE20 Five Easy	Mettler Toledo, Schwerzenbach, Switzerland
Pipetus-accu	Hirschmann, Neckartenlingen
Roller RM5 Assistant 348	Karl Hecht GmbH, Sondenheim
Scale model 1219 MP and H10T	Satorius, Göttingen
Shaker Rocky NG	Frögel Labortechnik, Lindau
Table top centrifuge, Biofuge pico	Heraeus, Hanau
Thermocylcer GeneAmp PCR System 9700	Applied Biosystems, Darmstadt
Thermomixer compact, heating block	Eppendorf, Hamburg
UV-table GelDoc 2000	BioRad, Munich
Vortex Genie2	Bender&Hobein, Zurich, Switzerland
Water bath	GFL, Burgwedel

2.1.2 Consumables

1.5 ml reaction tubes	Eppendorf, Hamburg
Cell culture 6 well-plates	Cornig, Amsterdam, Netherlands
Cell culture 96 well-plates	Cornig, Amsterdam, Netherlands
Cell culture dishes 10 cm	Greiner, Frickenhausen
Cell culture flasks	Sarstedt, Nümbrecht
Cell strainer 100 µm	Becton Dickinson, Heidelberg
Cryo tubes	Nunc, Denmark
FACS-tubes	Greiner, Frickenhausen
Fujifilm Medical XRAY Film Super RX	Fujifilm, Düsseldorf
Gloves	Rösner-Mauby Meditrade, Kiefersfelden
LS Columns	Miltenyi, Bergisch Gladbach
Nitrocellulose membrane hybond^{TM}	GE healthcare, Freiburg
Parafilm	Pechiney Plastic Packaging, USA
Pipette tips	Sarstedt, Nümbrecht
Pipette tips with filters	BioZym, Oldendorf
Pipettes	Becton Dickinson, Heidelberg
Polystyrene tubes	Becton Dickinson, Heidelberg
QIAshredder columns	Qiagen, Hilden
qRT-PCR 96 well plates	Roche, Mannheim
Reagent reservoirs	Cornig, Amsterdam, Netherlands
Sterile filter units (0.45 µm, 0.22 µm)	Millipore, Schwalbach
Syringe, 0.5 ml	Becton Dickinson, Heidelberg
Syringe, 10 ml	Servoprax, Wesel
Whatman paper 3MM	Whatman, Malstone, UK

2.1.3 Chemical reagents

4-hydroxy-3nitrophenyl acetate chicken gamma globuline (NP-CGG)	Biosearch Tech., Novato, USA
agarose UltraPure™	Invitrogen, Karlsruhe,
aluminium-potassium-dodecahydrate ($KAL(SO_4)_2$)	Merck, Darmstadt
ammoniumpersulfate (APS)	Carl Roth GmbH, Karlsruhe
Bovine serum albumine (BSA)	PAA Laboratories, Pasching, Austria
Bradford reagent RotiQuant®	Carl Roth GmbH, Karlsruhe
bromphenol blue	Fluka Chemie AG; Buchs, Switzerland
carboxyfluorescein diacetate succinimidyl ester (CFSE)	Invitrogen, Karlsruhe
citric acid	Sigma-Aldrich, Taufkirch
Concavalin A	Sigma-Aldrich, Taufkirch
deoxy nucleotide triphosphates (dNTPs)	GE healthcare, Uppsala, Sweden
dimethyl sulfoxide (DMSO)	Merck, Darmstadt
dithiothreitol (DTT)	Carl Roth GmbH, Karlsruhe
Dulbecco´s modified eagles medium (DMEM)	Invitrogen, Karlsruhe
ECL Western Blotting Reagents	GE Healthcare, Uppsala, Sweden
ethanol	Carl Roth GmbH, Karlsruhe
ethidiumbromide	Carl Roth GmbH, Karlsruhe
ethylendiamintetraacetate (EDTA)	Carl Roth GmbH, Karlsruhe

FasL-Flag, recombinant	Alexis Biochemical, San Diego, USA
fetal calf serum (FCS)	Sigma-Aldrich, Taufkirch
fluorescein digalactoside	Invitrogen, Karlsruhe
Forene® Isofluran	Abbott GmbH, Wiesbaden
formaldehyde [37%]	Carl Roth GmbH, Karlsruhe
glycerol	Carl Roth GmbH, Karlsruhe
Hank's balanced salt solution (HBSS)	Invitrogen, Karlsruhe
HEPES	Carl Roth GmbH, Karlsruhe
Histopaque®-1083 Ficoll	Sigma-Aldrich, Taufkirch
hydrogen peroxide (H_2O_2)	Sigma-Aldrich, Taufkirch
IL-2	Roche, Mannheim
ionomycine	Merck, Darmstadt
L-glutamine	PAA Laboratories, Pasching, Austria
lipopolysaccharide	Sigma-Aldrich, Taufkirch
magnesium chloride ($MgCl_2$)	Carl Roth GmbH, Karlsruhe
milk powder, low fat	AppliChem, Darmstadt
non-essential amino acids (100x)	Invitrogen, Karlsruhe
NP(17)-BSA	Biosearch Tech., Novato, USA
NP(3)-BSA	Biosearch Tech., Novato, USA
NP-40; IGEPAL CA-630	Sigma-Aldrich, Taufkirch
o-phenylenediamine tablets	Sigma-Aldrich, Taufkirch
penicillin/streptomycin	PAA Laboratories, Pasching, Austria
phenylmethylsulfonylfluoride (PMSF)	Sigma-Aldrich, Taufkirch
phorbol 12-myristate 13-acetate (PMA)	Sigma-Aldrich, Taufkirch
phosphate buffered saline (PBS; sterile)	PAA Laboratories, Pasching, Austria
polyethylenimine (PEI)	Sigma-Aldrich, Taufkirch
poly-L-lysine	R&D Systems, Wiesbaden
ponceau solution	Fluka Chemie AG; Buchs, Switzerland
potassium chloride (KCl)	Carl Roth GmbH, Karlsruhe
propidium iodide	Sigma-Aldrich, Taufkirch
Rosewell park memorial institute (RMPI)	Bio Whittacker, Belgium
Rotiphorese-Gel 30 (30% acrylamide/0.8% bisacrylamide)	Carl Roth GmbH, Karlsruhe
sodium pyruvat solution (100x)	Invitrogen, Karlsruhe
sodiumacetate (NaAc)	Carl Roth GmbH, Karlsruhe
sodiumazide (NaN3)	Carl Roth GmbH, Karlsruhe
sodiumcarbonate (Na_2CO_3)	Carl Roth GmbH, Karlsruhe
sodiumchlorid (NaCl)	Carl Roth GmbH, Karlsruhe
sodiumdodecylsulfate (SDS)	Carl Roth GmbH, Karlsruhe
sodiumhydrogenecarbonate (Na_2HCO_3)	Carl Roth GmbH, Karlsruhe
sodiumhydroxide (NaOH)	Carl Roth GmbH, Karlsruhe
sodiummcitrate (NaCi)	Carl Roth GmbH, Karlsruhe
Staphylococcus enterotoxin B	Sigma-Aldrich, Taufkirch
SYBR Green PCR master mix	Thermo Scientific, Bonn
tetramethylethylendiamine (TEMED)	Carl Roth GmbH, Karlsruhe
thioglycollate broth	Fluka Chemie AG; Buchs, Switzerland
tris(hydroxymethyl)-aminomethan (Tris)	Carl Roth GmbH, Karlsruhe
triton-X 100	Fluka Chemie AG, Buchs, Switzerland
Trypsin/EDTA	Invitrogen, Karlsruhe
tween-20	Carl Roth GmbH, Karlsruhe
β-mercaptoethanol	Carl Roth GmbH, Karlsruhe

2.1.4 Inhibitors

Complete Protease Inhibitor Cocktail	Roche, Mannheim
FasFc fusion protein, recombinant	Apogenix, Heidelberg
GF 109203x hydrochloride	Sigma-Aldrich, Taufkirch
Ly294002 hydrochloride	Sigma-Aldrich, Taufkirch
Phosphatase Inhibitor Cocktail 2	Sigma-Aldrich, Taufkirch
Raf1 Kinase Inhibitor I	Merck, Darmstadt
U-0126	Merck, Darmstadt
U-73122	Merck, Darmstadt

2.1.5 Enzymes

Platinum® *Pfx* DNA Polymerase	Invitrogen, Karlsruhe
RED*Taq*® Genomic DNA Polymerase	Sigma-Aldrich, Taufkirch
RNase A	Roche Diagnostics, Mannheim

2.1.6 Size standards

Benchmark™ Prestained Protein Ladder	Invitrogen, Karlsruhe
GeneRuler™ 1 kb PLUS DNA ladder	Fermentas GmbH, St. Leon Rot

2.1.7 Commercial kits

B cell isolation kit	Miltenyi, Bergisch Gladbach
BrdU Flow kit, FITC	Becton Dickinson, Heidelberg
$CD4^+CD25^+$ Regulatory T cell isolation kit	Miltenyi, Bergisch Gladbach
$CD4^+$ T cell isolation kit	Miltenyi, Bergisch Gladbach
$CD8^+$ isolation kit	Miltenyi, Bergisch Gladbach
Cell-Based ELISA base kit 3	R&D Systems, Wiesbaden
Cell-Based ELISA base kit 4	R&D Systems, Weisbaden
DNeasy Tissue kit	Qiagen, Hilden
ECL detection kit	GE healthcare, Freiburg
FASER kit, PE	Miltenyi, Bergisch Gladbach
Inside stain kit	Miltenyi, Bergisch Gladbach
Omniscript® Reverse transcriptase kit	Qiagen, Hilden
OptEIA™ Mouse TNF ELISA Kit II	BD Bioscience, Heidelberg
Pan T cell isolation kit	Miltenyi, Bergisch Gladbach
pAkt (S473) Pan Specific Cell-Based ELISA	R&D Systems, Wiesbaden
pERK1 (T202/Y204)/ERK2 (T185/Y187) Cell-Based ELISA	R&D Systems, Wiesbaden
RNase-free DNase set	Qiagen, Hilden
RNeasy® Mini kit	Qiagen, Hilden
RT^2 First strand kit	SABiosciences, Frederick, USA
RT^2 Profiler™ PCR Array Mouse Wnt Signaling Pathway	SABiosciences, Frederick, USA
RT^2 SYBR Green qPCR Master Mix	SABiosciences, Frederick, USA

2.1.8 Buffer and solutions
Cell biology

Annexin V binding buffer	10mM HEPES pH 7.4 150mM NaCl 5mM KCl 1mM $MgCl_2$ 1.8mM $CaCl_2$
FACS buffer	3% FCS/PBS
MACS buffer	0.5% BSA/2mM EDTA/PBS

ELISA

[0.2M] carbonate buffer pH 9.5 (for 1 l)	16.8 g $NaHCO_3$ 10.6 g Na_2CO_3 adjust to pH 9.5
ELISA-blocking buffer	1% milkpowder (low fat)/PBS
Fixation solution	8% formaldehyde/PBS
Quenching buffer	0.6% H_2O_2/PBS
Substrate buffer pH 5.0 (for 1 l)	10.5 g [0.1M] citric acid 12.1 g [0.1M] Tris adjust to pH 5.0

Molecular biology

Tris borate EDTA (TBE, 10x)	108.0 g Tris 55.0 g boric acid 40 ml [0.5M] EDTA pH 8.0 fill up to 1 l with H_2O
Phosphate buffered saline (PBS, 10x)	80.0 g NaCl 2.0 g KCl 14.4 g Na_2HPO_4 2.4 g KH_2HPO_4 fill up to 1 l with H_2O
Tris buffered saline (TBS, 10 x)	250mM Tris-HCl pH7.8 1.5M NaCl adjust pH to 8.1 fill up to 1 l with H_2O
DNA loading dye (6x)	50% (w/v) glycerol 1mM EDTA 0.04% bromphenolblue 0.4% (w/v) xylenecyanol
[0.5M] Tris-HCl, pH 6.8	30.1g Tris in 500 ml H_2O adjust pH to pH 6.8
[1.5M] Tris-HCl, pH 8.8	91.1 g Tris in 500 ml H_2O adjust pH to pH 8.8

Protein biochemistry

Blocking buffer	5% milkpowder (low fat) 1x PBS 0.05% Tween
Cell lysis buffer	10mM KCl 1.5mM $MgCl_2$ 10mM Tris-HCl pH 7.4 0.5% SDS 2% NP-40 1 tablet (in 50 ml) protease inhibitor cocktail
SDS-PAGE 2.5% stacking gel	2.5 ml Rotiphorese-Gel 30 3.7 ml [0.5M] Tris pH 6.8 8.5 ml H_2O 150 µl [10%] SDS 10 µl TEMED 100 µl [10%] APS
SDS-PAGE 12.5% running gel	4.2 ml Rotiphorese-Gel 30 2.5 ml [1.5 M] Tris, pH 8.8 3.2 ml H_2O 0.1 ml [10%] SDS 7 µl TEMED 70 µl [10%] APS
SDS-PAGE running buffer	1.5% [20% SDS] 300.0 g Tris 750.0 g glycerol H_2O up to 5 l
SDS-PAGE sample buffer (5x)	62.5mM Tris HCl pH 7.8 50mM DTT 20% glyerine 2% SDS 0.01% bromphenol blue
Semi-dry blotting buffer (10x)	291.0 g Tris 145.0 g glycerol H_2O up to 5 l
Semi-dry blotting buffer (working solution)	10% 10x stock solution 70% H_2O 20% methanol
Washing buffer	PBS/0.05% Tween

2.1.9 Antibodies

All antibodies used recognize murine proteins.

Cell stimulation

name	isotype	supplier
CD3ε (clone 145-2C11)	hamster	Becton Dickinson, Heidelberg
CD40 (clone FGK 45.5)	rat	Miltenyi, Bergisch Gladbach
IgM, µ chain-specific	rabbit	antibodies-online, Aachen

ELISA

name	isotype	supplier
IgG1, purified	mouse	Becton Dickinson, Heidelberg
IgG1-biotin	rat	Becton Dickinson, Heidelberg
IgM, purified	mouse	Becton Dickinson, Heidelberg
IgM-biotin	rat	Becton Dickinson, Heidelberg
MEK1 (H-8)	mouse	Santa Cruz, Heidelberg
pMEK1/2 (Ser218/Ser222)	goat	Santa Cruz, Heidelberg
pRaf-1 (Ser338)	goat	Santa Cruz, Heidelberg
Raf-1 (C20)	rabbit	Santa Cruz, Heidelberg
Streptavidin-AP	-	Vector Labs, Burlingame, USA

FACS

name	isotype	supplier
Alexa Fluor 546-anti goat	donkey	Molecular probes, Karlsruhe
Annexin V-APC	rat	Becton Dickinson, Heidelberg
B220-PE	rat	Becton Dickinson, Heidelberg
B220-PerCp	rat	Becton Dickinson, Heidelberg
CD138 (Syndecan)-biotin	rat	Becton Dickinson, Heidelberg
CD19-APC	rat	Becton Dickinson, Heidelberg
CD25-Pe	mouse	Miltenyi, Bergisch Gladbach
CD3ε-PE	rat	Becton Dickinson, Heidelberg
CD4-FITC	rat	Miltenyi, Bergisch Gladbach
CD4-PE	rat	Becton Dickinson, Heidelberg
CD4-TriColour	rat	Invitrogen, Karlsruhe
CD69-PeCy7	hamster	Becton Dickinson, Heidelberg
CD80-APC	hamster	Becton Dickinson, Heidelberg
CD86-FITC	rat	Becton Dickinson, Heidelberg
CD8α-FITC	rat	Becton Dickinson, Heidelberg
CD8α-TriColour	rat	Invitrogen, Karlsruhe
Fas (JO2)-PE	hamster	Becton Dickinson, Heidelberg
Fas (JO2)-PeCy7	hamster	Becton Dickinson, Heidelberg
FasL (MFL3)-PE	hamster	Becton Dickinson, Heidelberg
Fc blocking reagent, mouse	-	Miltenyi, Bergisch Gladbach
FoxP3-APC	mouse	Miltenyi, Bergisch Gladbach
IgD-FITC	rat	Becton Dickinson, Heidelberg
IgG1 (H+L)-PE (isotype control)	hamster	Becton Dickinson, Heidelberg
IgG1-FITC	rat	Becton Dickinson, Heidelberg
IgM-APC	rat	Becton Dickinson, Heidelberg
IgM-FITC	rat	Becton Dickinson, Heidelberg
IgM-PE	rat	Becton Dickinson, Heidelberg
Lef-1 (C19)	goat	Becton Dickinson, Heidelberg
Ly6G-FITC	rat	Becton Dickinson, Heidelberg
NP7-PE	-	Biosearch Tech., Novato, USA
PNA-FITC	-	Vector Labs, Burlingame, USA
pPKCα (pT497)-PE	mouse	Becton Dickinson, Heidelberg
pPLCγ2 (pY759)-Alexa Fluor 488	mouse	Becton Dickinson, Heidelberg
Straptavidin-PE	-	Becton Dickinson, Heidelberg
Vβ6-FITC	mouse	Becton Dickinson, Heidelberg
Vβ8-FITC	rat	Becton Dickinson, Heidelberg

Western blot

name	isotype	supplier
β-Actin	goat	Santa Cruz, Heidelberg
HRP-rabbit IgG	donkey	GE Healthcare, Freiburg
HRP-goat IgG	rabbit	Zymed, Vienna, Austria
pErk1/2 (Thr202/Tyr204)	rabbit	New England Biolabs, Frankfurt

2.1.10 Vectors

pEVR-FL9B Lef-1   Wiebke Baum

2.1.11 Oligonucleotides

Oligonucleotides were purchased from BioSpring (Frankfurt) and resolved in HPLC-H$_2$O at a concentration of 100µM. Sequences are shown in 5' → 3' orientation.

PCR primer for genotyping of mice using genomic DNA

gene	primer	sequence	product size
fasl	loxP-ctrl-for	CCATGAAATCATTTTATGGTTGGGG	wt: 219 bp
	loxP-ctrl-rev	GAACAAGACACAAAATTGATTTTCATC	ΔIntra: 280 bp
fas^{lpr}	lpr-common	GTA AAT AAT TGT GCT TCG TCA G	
	lpr-wildtype	CAA ATC TAG GCA TTA ACA GTG	wt: 179 bp
	lpr-mutant	TAG AAA GGT GCA CGG GTG TG	Fas^{lpr}: 217 bp
lacZ	lacZ-for	CGGTGATGGTGCTGCGTTGGA	370 bp
	lacZ-rev	ACCACCGCACGATAGAGATTC	

PCR primer for detection of mRNA transcripts (semi-quantitative)

gene	primer	sequence	product size
fasl	screen-for	GCGGAAACTTTATAAAGAAAACTTAG	wt: 460 bp
	screen-rev	CGGAGTTCTGCCAGTTCCTTCTG	ΔIntra: 220 bp
lef-1	Lef1-for	TGCAGCTATCAACCAGATCC	311 bp
	Lef1-rev	GGGTTTCAACAAGCTTCCAT	

qRT-PCR primer

gene	forward primer	reverse primer
survivin	AGGCAAAGGAGACCAACAAC	ATGTGGCATGTCACTCAGGT
ccl 3	TCTCCTACAGCCGGAAGATT	CTCTCAGGCATTCAGTTCCA
cyclin D2	ATTTCAAGTGCGTGCAGAAG	AGGTAATTCATGGCCAGAGG
c-myc	GTGTCTGTGGAGAAGAGGCA	GCGTAGTTGTGCTGGTGAGT
cxcl 10	CTGAGTGGGACTCAAGGGAT	TTTCATCGTGGCAATGATCT
cyclin D1	GCCGAGAAGTTGATGCATCTA	TGGTCTGCTTGTTCTCATCC
eda	GGGCCTCAGATAGTGGTTGT	CAAAGAAACCCAGGAAGAGC
egr 1	GTTATCCCAGCCAAACGACT	ATGAAGAGGTCGGAGGATTG
fgf 4	ACGAGGGACAGTCTTCTGGA	AAGAAAGGCACACCGAAGAG
gadda	AGAGCAGAAGACCGAAAGGA	GCACAGTACCACGTTATCGG
gapdh	TGGCAAAGTGGAGATTGTTG	CATTATCGGCCTTGACTGTG
histone1, h4a	CGTGGTAAGGGTGGTAAAGG	GATCACGTTCTCCAGGAACA
hprt 1	TCCTCCTCAGACCGCTTTT	CCTGGTTCATCATCGCTAATC
il-10	CCCAGAAATCAAGGAGCATT	TCACTCTTCACCTGCTCCAC
il-4	ACGGCACAGAGCTATTGATG	TGCCGATGATCTCTCTCAAG
il-5	TGAAAGAGACCTTGGCACTG	CTCCAGTGTGCCTATTCCCT
irf 4	GAGCTGTAGGTTGCCCAGAT	AGAAGAGGGAGCAAACAGGA
jak 1	TGAACCACCTCAAGAAGCAG	GGGCTTTCTTAGTGGCTACG
lef-1	TGCAGCTATCAACCAGATCC	GGGTTTCAACAAGCTTCCAT
nfatc 1	CCAGTATACCAGCTCTGCCA	GTGGGAAGTCAGAAGTGGGT
nfkb	GAAGGCTGGAGAAGATGAGG	CCAGCAGTTTGCAGAGTTGT
nr4a1	CCTGGACGTTATCCGAAAGT	AAGAGTTCCAGGAAGGCAGA
numb	TGGCAGACAGATGAAGAAGG	CCTTCACTGCTTTCTTTCCC
phlda1	GAAGGAAGGAGTGCTGGAAA	CCACTAGGTTGGGACGGTT
s100a6	GGAGCTGAAGGAGTTGATCC	AGGCGACATACTCCTGGAAG
sprouty2	CATCAGAGGCAGCAATGAAT	TGCTATTCACATTGGCTGGT
tagap	CCATTGTCCAACAAGACAGG	ATCTTCAGCTTCATCGCCTT
tnfα	GAGCACAGAAAGCATGATCC	CCACAAGCAGGAATGAGAAG

2.1.12 Mouse lines

All mice were bred to a pure black-six background.

name	description	reference/origin
FasL ΔIntra	'knockout/knockin' of wildtype *FasL* with a truncated *FasL* lacking aa 2-74	Group of PD Dr. M. Zörnig
B6Smn.C3-*FasL^{gld}*/J (*gld*)	A-to-C point mutation near the 3' end of the coding sequence resulting in an aa exchange (F273L)	Jackson laboratory
FasB6.MRL-*Fas^{lpr}*/J (*lpr*)	transposon insertion in *Fas* intron 2	Jackson laboratory
FasL knockout (*FasL$^{-/-}$*)	knockout of the complete *FasL* locus	(Karray *et al.*, 2004)
FasL ΔIntra/Fas^{lpr}	cross of heterozygous *FasL ΔIntra* mice with *Fas^{lpr}* mice	-
FasL ΔIntra/TopGaL	cross of heterozygous *FasL ΔIntra* mice with *TopGal* mice (Stock Tg(Fos-lacZ)34Efu/J)	*TopGal* mice (DasGupta *et al.*, 1999) obtained from Dr. S. Liebner, University Frankfurt

2.2 Methods

2.2.1 Animal models

Mice were bred and maintained under specific pathogen-free conditions in filter top- or isolated, ventilated cages (IVC) according to the German animal welfare law. For experimental use, mice were anaesthetized with isofluran and sacrificed by cervical dislocation. All mice were bred to a pure C57Bl/6N background.

2.2.2 Genotyping

The genotype of mice was determined by amplifiying a fragment of the respective targeted locus *via* PCR. For this purpose, genomic DNA was isolated from tail biopsies using the DNeasy Tissue Kit according to the manufacturer's instructions (*Qiagen*). The tissue was digested in the presence of 20 µl Proteinase K and 180 µl lysis buffer at 55°C overnight. Using spin column purification, DNA was bound to the column, repeatedly washed and eluted with 200 µl elution buffer. Purified genomic DNA was stored at 4°C until further use. PCR was performed on 2 µl genomic DNA using Red*Taq* genomic DNA Polymerase (S*igma*) as follows:

PCR reaction mix		Thermo cycler program		
		temperature [°C]	time [min]	cycles
tail DNA	2.0 µl	94	3	1
dNTP mix [2.5nM each]	3.0 µl	94	1	
5' primer [3µM]	5.0 µl	57	1	30
3' primer [3µM]	5.0 µl	72	3	
10x *Taq* buffer	2.5 µl	72	10	1
Red*Taq* gDNA Polymerase	1.5 µl	4	∞	1
H$_2$O	6.0 µl			

PCR products were analyzed on a 2% agarose-gel.

For genotyping of the *Fas^{lpr}* locus, a three primer strategy was employed using primers at 20µM each and annealing at 59°C according to the protocol provided by the Jackson Laboratory.

2.2.3 Cell culture methods

2.2.3.1 Isolation of naïve murine lymphocytes

Primary lymphocytes were isolated by negative selection using magnetic bead purification and MACS technology according to the manufacturer's instructions (*Miltenyi*). Negative selection allows the purification of untouched cells of interest. The appropriate isolation kits contain mixtures of biotinylated antibodies recognizing lineage markers of unwanted cell populations. Antibody-cell complexes are coupled to magnetic micro beads which are retained on the purification column when placed in a magnetic field, while the cells of interest are in the flow through.

Mice were sacrificed and the spleen was removed, transferred into 10 ml cold PBS and grinded through a screen cup sieve using a syringe plunger. Erythrocytes in the homogenate were lyzed by hypotonic cell lysis. To allow antibody-cell complex formation, cell suspensions were incubated with 40 µl MACS buffer and 10 µl biotinylated antibody mix per 10^7 cells for ten minutes at 4°C. Biotinylated antibodies were coupled to magnetic micro beads in the presence of additional 30 µl MACS buffer and 20 µl anti-biotin magnetic microbeads per 10^7 cells for 15 minutes at 4°C. After washing with ten times of the volume of the labeling reaction, the pellet was resuspended in 500 µl MACS buffer/10^8 cells and applied onto an equilibrated MACS LS-column placed in a MACS magnetic field. Columns were allowed to empty by gravity flow and were subsequently washed three times by the addition of 3 ml MACS buffer each. Cell number and viability were analyzed with a Casy counter. Purity of the cell population was determined by flow cytometry using appropriate lineage surface markers (B cells: CD19 or B220; T cells: CD4 and CD8).

2.2.3.2 Generation of T cell blasts

For the generation of T cell blasts, spleen homogenates were incubated at a concentration of 3×10^6 cells/ml for two days in the presence of 2 µg/ml Concavalin A. After repeated washing, cells (3×10^6 cells/ml) were cultivated for further five days in the presence of 100 U/ml human-recombinant IL-2. Fresh medium was added when required. For the final isolation of viable T cells, samples were subjected to density gradient centrifugation. The cell suspension was underlayed with a Ficoll solution (density 1.083) 1:1 and centrifuged at 800 xg for 20 minutes without brake. The interphase cells were collected, washed twice with PBS and used for co-culture experiments.

2.2.3.3 Lymphocyte maintenance and stimulation

Primary lymphocytes were maintained in Rosewell Park Memorial institute medium (RMPI 1640) medium supplemented with 10% FCS, 2% L-glutamine, 1% penicillin/streptomycin, 1% sodium pyruvat solution (100x), 1% non-essential amino acids (100x) and 0.1% β-mercaptoethanol (1000x) at 37°C and 5% CO_2.

Stimuli used to activate B cells include various concentrations of solubly added anti-IgM antibody, anti-CD40 antibody, lipopolysaccharide (LPS) or 100 ng/ml phorbol myristate acetate (PMA)/0.2 µg/ml ionomycine. T cells were activated using various concentrations of plate-bound anti-CD3ε antibody or 100 ng/ml PMA/0.2 µg/ml ionomycine.

2.2.3.4 Erythrocyte lysis

Prior to lymphocyte isolation from lymphoid organs and to FACS analysis of organ homogenates, erythrocytes were lyzed by hyptonic cell lysis. Cell suspensions were pelleted by centrifugation at 500 xg for five minutes, resuspended in 500 µl 1x PBS and lyzed by the addition of 5 ml H_2O for ten seconds. The reaction was stopped with 500 µl 10x PBS and then 5 ml 1x PBS. Removal of cell debris was achieved by passing the cell suspension through a 100 µm cell strainer.

2.2.3.5 Determination of cell number and viability

The total cell number and the number of viable cells in a sample were determined with a Casy cell counter. The Casy technology exploits the property of intact cell membranes to function as an insulator for electric currents. An appropriate programm for measurement was established using Casyblue, following the manufacturer's instructions.

Casy counter settings

dilution of cell suspension	1:400
measured sample volume	400 µl
measurement cycles	3
auto aggregation	on
normalization cursor	3.2 µm
evaluation cursor	5.3 µm

2.2.3.6 Cultivation of cell lines

All cells were maintained at 37°C and 5% CO_2 in humid atmosphere in an incubator. The murine B cell lymphoma cell line A20 (ATCC number TIB-208) was maintained in RMPI medium containing 10% FCS, 2% L-gluatmine, 1% penicillin/streptomycin and 0.1% β-mercaptoethanol. Cells were split 1:10 approximately every third day. The adherent human epithelial kidney cell line HEK293T, which was derived from the HEK293 cell line (ATCC number CRL11268) by transformation with the SV40 large T-antigen, was cultivated in DMEM supplemented with 10% FCS, 2% L-glutamine and 10% penicillin/streptomycin. For cell passaging, growth medium was removed and cells were washed with PBS. Addition of 1 ml Trypsin-EDTA facilitated cell dispersal. The Trypsin reaction was stopped with growth medium, and cells were resuspended by pipetting and split 1:10.

2.2.3.7 Freezing and thawing of cells

Cells were pelleted and resuspended in freezing medium (10% DMSO/20% FCS/growth medium without antibiotics). The suspension was slowly cooled to -80°C by placing the vials into an isopropanol filled box (cooling rate 1°C/min) and subsequently transferred to liquid nitrogen for long term storage. Frozen cells were swiftly thawed at 37°C and mixed with 5 ml complete growth medium. Cells were then pelleted to remove DMSO, resuspended in growth medium and cultivated as desired.

2.2.3.8 Transfection of cells

HEK293T cells were transiently transfected using polyethylenimine (PEI). One to two million cells were seeded onto a 10 cm cell culture dish to achieve a confluencey of ~60% at the day of transfection. Next day, the transfection reaction was assembled by mixing 5 µg plasmid DNA, 13.5 µl 100mM PEI and 300 µl PBS and incubating this mixture for 15 minutes at room temperature to allow DNA-PEI complex formation. The culture medium of cells was exchanged for growth medium without antibiotics. Subsequently, the transfection mix was added dropwise and distributed evenly by gently swaying the culture dish. Four hours later, medium was supplemented with FCS to a concentration of 10%. Cells usually were harvested 48 hours post-transfection.

2.2.4 Flow cytometry

2.2.4.1 Analysis of cell surface marker expression

Single cell suspensions of purified lymphocytes or lymphoid organ homogenates were prepared, pelleted, resuspended in 100 µl FACS buffer and stained for cell surface molecules for 20-30 minutes at 4°C. Fluorochrome-conjugated antibodies recognizing FasL (MFL3), Fas (JO2), CD3ε, CD4, CD8α, CD19, B220, CD69, CD80, CD86, IgM, IgD, IgG1, CD138 (Syndecan), Peanut agglutinin (PNA), Vβ6 TCR chain, Vβ8 TCR chain, CD11b, Ly6G (Gr-1), F4/80, Annexin V and isotype control antibodies were purchased from *BD Bioscience*. For all stainings, cells were pre-incubated with FcBlock (1:10; *Miltenyi*) for ten minutes at room temperature to block unspecific binding of antibodies to Fc receptors present on lymphocyte cell surfaces. Propidium iodide (2.5 µg per sample) was used for live/dead cell discrimination wherever required. At least 10,000 cells per sample were analyzed using a FACSCalibur with the professional CellQuest software for evaluation.

2.2.4.2 Intracellular stainings

Intracellular stainings were performed using the Inside Stain Kit from *Miltenyi*, including the reagents InsidePerm and InsideFix for cell permeabilization and fixation, according to the manufacturer's instructions. All incubations were performed at room temperature. Cells were resupended in 250 µl FACS buffer per 10^6 cells and fixed with 250 µl/10^6 cells InsideFix (containing 3.7% formaldehyde) for 20 minutes, prior to staining with 100 µl of the desired antibody diluted in InsidePerm for 20 minutes. After washing with 1 ml InsidePerm, cells were resuspended in 300 µl FACS buffer, and at least 10,000 cells per sample were analyzed using a FACSCalibur (*BD Bioscience*). Lef-1 protein was detected with a goat anti-mouse Lef1 antibody (C19) in a dilution of 1:50 and an Alexa Fluor546-coupled anti-goat antibody (1:250) as secondary antibody. For the analysis of phosphorylated proteins in primary lymphocytes cells were rested for one hour after isolation before they were activated as desired. Labeled antibodies to phosphorylated PLCγ2 and phosphorylated PKCα were purchased from *BD Bioscience* and used according to the manufacturer's instructions (15 µl per test).

2.2.4.3 Staining of unfixed cells with Annexin V and propidium iodide

Spontaneous and activation-induced cell death was assessed by staining with Annexin V and propidium iodide. Annexin V binds to phosphatidylserine exposed on the surface of apoptotic cells. Propidium iodide discriminates live and dead cells, as it is excluded from intact cells while diffusing into dead cells with a porous membrane.

Cells were washed with Annexin V binding buffer and stained with 3 µl Annexin V-APC diluted in 200 µl Annexin V binding buffer for 15 minutes at room temperature. Before flow cytometric measurement, 2.5 µg propidium iodide were added. Late apoptotic and necrotic cells stain positive for Annexin V and propidium iodide, while early apoptotic cells become Annexin V positive and PI negative.

2.2.4.4 Propidium iodide staining of ethanol fixed cells – Nicoletti test

Nicoletti *et al.* (Nicoletti *et al.*, 1991; Riccardi *et al.*, 2006) described a method how to simultaneously quantify the number of living and dead cells in one sample and to analyze their cell cycle status. After fixation and permeabilization, propidium iodide is used to stain cellular DNA. In apoptotic cells DNA is hydrolyzed and diffuses out of the cell. As propidium iodide intercalates into double-stranded DNA, the resulting fluoerescent staining is proportional to the ploidy of the cell.

Cells were fixed by dropwise addition of 1 ml ice-cold 70% ethanol and subsequent storage at 4°C overnight. After ethanol removal by washing with 38mM sodium citrate (pH 7.4), cells were stained with 50 µg/ml propidium iodide and 5 µg/ml RNase A in 38mM sodium citrate (pH 7.4) for 20 minutes at room temperature. The fluorescence light intensity in the FL-2 channel was quanitfied using the doublet discrimination module (DDM). A gate was set on the cell population present in a FL2-A/FL2-W dot plot, so that the cell cycle profile could be visualized in a FL2-A histrogram. The extent of cell death is represented by the percentage of cells in the sub-G1 peak of the cell cycle profile (hypoploid cells).

2.2.4.5 Carboxyfluorescein diacetate succinimidyl ester (CFSE) dilution assay for lymphocyte proliferation

Proliferation of lymphocytes can be tracked by staining of cells with the cell permeable dye CFSE. Taken up by living cells, it is cleaved by endogenous esterases to become cell impermeable and fluorescent. Each cell division diminishes the fluorescence by half, so that the decrease in staining intensity is proportional to cell proliferation.

Lymphocytes were incubated with 1μM CFSE in 3% FCS/PBS for three minutes at room temperature, washed twice and seeded at a density of 5×10^5 cells/100 μl growth medium in a 96well-plate. Cells were left untreated or were stimulated using various concentrations of plate-bound anti-CD3ε antibody (T cells), or soluble anti-CD40 antibody or rabbit anti-IgM antibody (B cells). 48 or 72 hours later cells were harvested and resuspended in 300 μl FACS buffer containing 2.5 μg propidium iodide. Proliferation of living cells was measured by FACS in the FL-1 channel as the decrease in CFSE fluorescence intensity.

2.2.4.6 Co-culture of primary lymphocytes with Fas-sensitive A20 target cells

To assess the capability of primary murine lymphocytes to kill target cells in a FasL/Fas-dependent manner, T cell blasts were co-cultured with Fas-expressing A20 target cells. T cell blasts were re-stimulated with 10 μg/ml plate-bound anti-CD3ε antibody for 24 hours. Afterwards, half of each cell population was pre-incubated with 100 μg/ml FasFc for 30 minutes at 37°C to specifically block FasL-dependent killing. A20 target cells were labeled with CFSE. Activated T cells, pre-incubated with FasFc or left untreated, and labeled A20 cells were co-cultured in an effector:target ratio of 2.5:1 for six hours in the presence of 10 μg/ml anti-CD3ε antibody. Cells were then fixed with ice-cold 70% ethanol and stained with propidium iodide. FasL-dependent killing was quantified by substracting the percentage of cells with hypoploid DNA content (sub-G1 cells) of FasFc-treated samples from that of sub-G1 cells in the corresponding untreated sample. Experimental controls included A20 cells not subjected to co-culture but incubated in the presence of recombinant FasL protein (67 ng/ml FasL-Flag/1 μg/ml anti-Flag antibody (M2)), recombinant FasL protein plus 100 μg/ml FasFc or FasFc only.

2.2.4.7 β-Galactosidase activity assay for the analysis of Wnt signaling *in vivo*

A *TopGal* based mouse line was used to analyze a potential influence of FasL reverse signaling on Wnt/Lef-1 signaling *in vivo* (DasGupta *et al.*, 1999; Barolo, 2006). *TopGal* mice express a β-Galactosidase reporter gene under the control of a Lef-1-dependent regulatory sequence, consisting of three consensus Lef-1 binding sites upstream of a minimal *c-Fos* promoter. Breeding with heterozygous mice of the *FasL ΔIntra* line resulted in offspring homozygous for either the wildtype or the mutant *FasL* locus and heterozygous for the *LacZ* locus. FACS analysis of β-Galactosidase-mediated sequential hydrolysis of Fluorescein Digalactoside (FDG) to Fluorescein Monogalactoside (FMG) and then to highly fluorescence

Fluorescein was performed in B cells derived form *FasL ΔIntra/TopGal* mice. Cells (1×10^6 cells/100 µl medium) were stimulated for 4-48 hours with 5 µg/ml anti-IgM antibody, 10 µg/ml anti-CD40 antibody, 10mM LiCl or were left untreated. Cells were harvested, and 50 µl cell suspension were transferred to a FACS tube. A 20mM FDG stock solution in DMSO was prepared and diluted tenfold in H_2O for use. Equal amounts of diluted FDG solution and cell suspension were mixed and incubated for one minute at 37°C for hypotonic cell lysis. The reaction was stopped with a tenfold volume ice-cold growth medium. During a subsequent incubation for 30-45 minutes on ice, any β-Galactosidase present in the sample converts FDG into its fluorescence form. For FACS analysis, 2.5 µg PI were added and β-Galactosidase activity was quantified in the FL-1 channel. B cells derived from Rosa mice were included as a positive control for the assay.

2.2.5 Enzyme-linked immunoabsorbant assay (ELISA)

2.2.5.1 Cell-based ELISA

Cell-based ELISAs are modified solid phase sandwich ELISAs. For analysis, cells are grown directly on the ELISA plate. Following treatment, cells are fixed, permeabilized and simultaneously incubated with primary antibodies recognizing two different epitopes (phosphorylated protein or total protein). The primary antibodies are derived from different species so that two appropriate enzyme-linked secondary antibodies can be used for the simultaneous detection of fluorgenic substrate conversion. Relative levels of phosphorylated protein are calculated by normalization to levels of the total protein. In this study, commerically available cell-based ELISAs for the detection of phosphorylated ERK1/2 (T202/Y204) and Akt (S473) were used according to the manufacturer's instructions (*R&D Systems*).

Phosphorylation of c-Raf and MEK1/2 was analyzed using the cell-based ELISA kits number 3 (MEK1/2) and number 4 (c-Raf) (*R&D Systems*). These ELISA kits follow the same protocol, but only contain two different enzyme-linked secondary antibodies. Primary antibodies against the proteins of interest can be chosen by the investigator. For the detection of c-Raf activation, goat anti-pRaf1 (S338) and rabbit anti-Raf1 (C20) antibodies were used (*Santa Cruz*). Total MEK1/2 protein and the phosphorylated isoforms were detected with mouse anti-MEK1 (H-8) and goat anti-pMEK1/2 (S218/S222) antibodies (*Santa Cruz*). Antibodies against c-Raf and MEK1/2 were used at a 1:50 dilution.

All incubation and washing steps were performed at room temperature with gentle shaking. Inbetween each consecutive assay step, samples were washed three times with 200 µl/well 1x wash buffer for five minutes each time unless stated otherwise. The substrates F1 and F2 were included in the kits (*R&D Systems*).

ELISA plates were coated with 100 µl/well poly-L-lysine (10 µg/ml) to promote lymphocyte adhesion. After incubation for 1.5 hours at 37°C, the poly-L-lysine solution was removed and plates were washed twice with PBS. Splenic B cells were isolated and 10^6 cells/100 µl medium were seeded onto the coated plate. Cells were allowed to recover from stress caused by the isolation procedure, that might lead to overall higher protein phosphorylation levels, by resting them for two hours. B cells were then stimulated by the addition of 1 µg/ml anti-IgM antibody for the indicated times, after which the reaction was stopped with 100 µl/well formaldehyde (8%). To properly fix the cells, plates were incubated for 20 minutes. Following incubation with 100 µl/well quenching buffer for 20 minutes, unspecific binding sites were blocked with 100 µl/well blocking buffer for one hour. Primary antibodies (100 µl/well) were allowed to bind overnight at 4°C. No primary antibody was added to negative control wells. The following day, the two secondary antibodies were applied for two hours (100 µl/well). For fluorgenic detection, cells were washed twice with 200 µl/well 1x wash buffer and a third time with 200 µl/well PBS. Thereafter, 75 µl/well substrate F1 were added for one hour. Finally, 75 µl/well substrate F2 were applied for 30-40 minutes, before measuring the fluorescence light intensity with an ELISA reader. Phosphorylated protein was detected by setting the excitation wavelength to 540 nm and that of the emission to 600 nm; total protein was detected by setting the excitation wavelength to 360 nm and that of the emission to 450 nm. For evaluation, background fluorescence was substracted from all samples. The relative amount of phosphorylated protein in each sample was calculated by dividing the fluorescence reading representing the phosphorylated isoform by that of the total protein.

2.2.5.2 Standard solid phase sandwich ELISA

Standard solid phase sandwich ELISAs detect proteins in cell culture supernatants, serum or plasma. Usually, assay plates are pre-coated with a specific antibody, so that the protein of interest in the sample binds to the immobilized antibody. The addition of a secondary enzyme-linked antibody produces an 'antibody sandwich'. Substrate conversion by the linked enzyme is directly proportional to the amount of protein of interest present in the sample and can be measured as the light absorbance by or the fluorescence intensity of the colored product in an ELISA reader. Here, the activation-induced TNFα secretion was quantified with a commerically available assay following the manufacturer's guidelines (*BD Bioscience*). Assay reagents were brought to room temperature before use. All incubations were done at room temperature with gentle shaking. Plates were washed inbetween consecutive assay steps by twenty times filling and decanting of wells with 300 µl/well 1x wash buffer.

B cells were isolated, seeded at a concentration of 10^6 cells/100 µl medium onto a 96 well-plate and were rested for two hours before stimulation with 1 µg/ml anti-IgM antibody for four, six or 24 hours. After the indicated times, cells were harvested from the plate, transferred to 1.5 ml reaction tubes and pelleted by centrifugation at 15,000 xg for five minutes at 4°C. Supernatant was collected and stored at -80°C until analysis. A standard containing a defined amount of recombinant mouse TNFα protein was serially diluted in Standard/Sample diluent to cover a concentration range between zero and 2000 pg/ml. After pipetting 50 µl ELISA diluent to wells pre-coated with an anti-TNFα antibody, 50 µl of each standard and sample were added in duplicate for two hours. After incubation with 100 µl/well biotinylated anti-TNFα detection antibody for one hour, 100 µl/well enzyme working reagent were applied for 30 minutes. In a final wash step, washing buffer was left in the wells for 30 seconds before decanting. Thereafter, 100 µl/well TMB one-step substrate reagent were added for 30 minutes. Plates were covered to protect from direct light. Substrate conversion was stopped by 50 µl/well Stop solution, and the absorbance was measured at 450 nm in an ELISA reader. For evaluation, the mean absorbance of each duplicate was calculated and the mean zero standard absorbance (background) was substracted. A standard curve was plotted with the TNFα concentration represented by the x-axis and the absorbance by the y-axis. The TNFα concentration in the samples was determined based on the linear regression of the standard curve.

2.2.6 Molecular biological methods

2.2.6.1 Isolation of total RNA

Total RNA was isolated with the RNeasy Mini Kit following the manufacturer's guidelines (*Qiagen*). For cell lysis, pelleted cells were resuspended in 600 µl lysis buffer RLT (containing β-mercaptoethanol), pipetted onto a QIAshredder column and centrifuged for two minutes at 15,000 xg. Lysates were mixed with one volume 70% ethanol, and the RNA was bound to RNeasy mini columns by centrifugation for 15 seconds at 15,000 xg. To eliminate contaminating genomic DNA, on-column digestion of genomic DNA with the RNase-free DNase set (*Qiagen*) was performed by adding 10 µl DNase I and 70 µl RL-buffer to the spin column membrane and incubating samples for 15 minutes at room temperature. After washing the column twice with 500 µl buffer RPE each, RNA was eluted with 42 µl HPLC-H_2O. RNA concentration was determined spectrophotometrically using a Nanodrop device.

2.2.6.2 Quantification of nucleic acid concentration and purity

Nucleic acids were diluted in water and the concentration was determined using a Nanodrop device based on the optical density readings at 260 nm and 280 nm. The quotient $OD_{260\,nm}/OD_{280\,nm}$ was used to determine the purity of nucleic acids. Purified RNA should display a ratio of 1.8 and DNA of 2.0 (Maniatis *et al.*, 1982).

2.2.6.3 Complementary DNA (cDNA) synthesis

cDNA synthesis using the Omniscript Reverse Transcriptase kit for First-Strand cDNA synthesis (*Qiagen*) was performed as described in the manual using 0.5-2 µg RNA. The RT-PCR reaction was set up to contain 1x RT buffer, 0.5mM of each dNTP, 1µM oligo(dT) primer, 10 units RNase inhibitor, 4 units Omniscript Reverse Transcriptase and H_2O up to a final volume of 20 µl. Reverse transcription proceeded at 37°C for 60 minutes in a thermo cycler. cDNA was stored at -20°C for further use.

2.2.6.4 Polymerase chain reaction (PCR)

PCR reactions were performed using 10 ng template cDNA derived from primary lymphocytes or transiently transfected HEK293T cells as follows:

PCR reaction mix		Thermo cycler program		
template cDNA	10 ng	temperature [°C]	time [min]	cycles
10x PCR buffer	5.0 µl	94	3	1
MgSo$_4$	2.0 µl	94	1	
dNTPs [2.5mM each]	6.0 µl	57	1	30
primer 5' [10µM]	1.5 µl	68	3	
primer 3' [10µM]	1.5 µl	68	10	1
Platinum *Pfx* Polymerase	0.8 µl	4	∞	1
H$_2$O up to 25 µl				

2.2.6.5 Electrophoresis of PCR products

PCR products were separated at 90V on a 2% agarose-gel (in 0.5x TBE) containing 0.1 µg/ml ethidiumbromide. Either the 100 bp DNA ladder or the 1 kb DNA ladder from Invitrogen was used as size standard.

2.2.6.6 Quantitative real time polymerase chain reaction (qRT-PCR)

qRT-PCR is an amplification method for nucleic acids based on standard PCR analysis but allowing the relative or absolute quantification of mRNA transcripts by fluorescence labeling of newly synthesized products. Here, relative quantification of mRNA transcripts was performed using cDNA derived from activated B cells (1 µg/ml anti-IgM antibody for four hours) and SYBR Green PCR Master Mix (*Thermo scientific*) with a Light Cycler 480 (*Roche*). One to two microgramms of RNA were transcribed into cDNA and for each pair of primers the reaction was set up in triplicates as follows:

PCR reaction mix		
component	one reaction [µl]	one sample (triplicate) [µl]
cDNA	0.67	2.00
5' primer [10µM]	1.25	3.75
3' primer [10µM]	1.25	3.75
SYBR Green PCR mix	12.50	37.50
H$_2$O	10.00	30.00

Light cycler program

temperature [°C]	time [min]	ramp rate [°C/sec]	Cycles
95	15:00	4.4	1
95	00:20	4.4	
57	00:30	2.2	40
72	00:30	4.4	
50	00:10	2.2	1
95	continuous	0.06	10 acquisitions per °C
37	∞		1

Each sample was assessed in triplicate and normalized to the expression levels of the housekeeping genes *Hypoxanthin-guanin-phosphoribosyltransferase* (*Hprt*) or *Gylerinaledhyde 3-phosphate dehydrogenase* (*Gapdh*). Primers were designed using the Genescript online resource (https://www.genscript.com/ssl-bin/app/primer), and specificity of their annealing was ensured by agarose-gel electrophoresis. To exclude a contamination in the PCR reaction, a triplicate sample containing water instead of cDNA was included for each set of primers used. Relative quantification of target gene expression was determined with the comparative threshold cycle method. For this purpose, the threshold cycle (C_t) value of the reference housekeeping gene was substracted from the C_t of the target gene to obtain the ΔC_t value. To relate the expression of a target gene in the study group (here: *FasL ΔIntra*) to that in the control group (here: wildtype) the $\Delta\Delta C_t$ value is calculated by substracting the respective ΔC_ts. Subsequent logarithmization yields the fold difference of mRNA levels. As the gene expression level for each target gene in the control group is set to one, values below one indicate reduced and those larger than one higher mRNA levels in the study group compared to the control group.

Exemplary calculation

Calculate the ΔC_t of each target gene in wildtype (Wt) and *FasL ΔIntra* (*ΔIntra*) samples

$\Delta C_t = C_t$ target gene/Wt $- C_t$ reference gene/Wt

$\Delta C_t = C_t$ target gene/*ΔIntra* $- C_t$ reference gene/*ΔIntra*

Relate the expression of a target gene in *FasL ΔIntra* mice to that in wildtype mice by calculating the $\Delta\Delta C_t$

Wildtype: $\Delta\Delta C_t = \Delta C_t$ Wt $- \Delta C_t$ Wt $= 0$

FasL ΔIntra: $\Delta\Delta C_t = \Delta C_t$ *ΔIntra* - ΔC_t Wt = relative difference in target gene expression level
Logarithmize the $\Delta\Delta C_t$ to the basis of two to determine the fold change in target gene expression level. Using a logarithmic basis of two assumes an optimal PCR efficient, where each product present is doubled in every cycle.
Wildtype: $2^{(-\Delta\Delta Ct)} = 2^{(0)} = 1$

FasL ΔIntra: $2^{(-\Delta\Delta Ct)}$ = fold difference in target gene mRNA levels in *FasL ΔIntra* cells compared to wildtype cells

2.2.6.7 mRNA expression profiling of Wnt signaling-related genes

For the detailed analysis of Wnt signaling in activated B cells the RT² Profiler™ PCR Array Mouse Wnt Signaling Pathway (*SABioscience*) was used according to the manufacturer's instructions. The array contains primers for 86 Wnt singaling pathway-related genes, housekeeping genes, probes to detect contaminating genomic DNA and a positive PCR control. RNA isolated from primary B cells that had been activated for four hours in the presence of 1 µg/ml anti-IgM antibody was transcribed into cDNA using the RT² First strand kit (*SABioscience*): Genomic DNA was eliminated by incubating 0.5 µg RNA (8 µl) and 2 µl GE (5x genomic DNA elimination buffer) at 42°C for five minutes, followed by chilling on ice for some minutes. For reverse transcription, a cocktail containing 4 µl BC3 (5x RT buffer 3), 1 µl P2 (primer and external control mix), 2 µl RE3 (Reverse Transcriptase enzyme mix 3) and 3 µl H₂O was prepared. Equal amounts of RNA solution and Reverse Transcriptase cocktail were mixed, incubated for exactly 15 minutes at 42°C, and the reaction was stopped by heating at 95°C for five minutes. After diluting the synthesized cDNA with 91 µl H₂O it was used together with the RT² SYBR Green qPCR Master Mix (*SABioscience*) for qRT-PCR. The reaction was assembeled to contain 102 µl cDNA, 1350 µl 2x RT² qPCR Master Mix and 1248 µl H₂O. Twentyfive microliters of this mixture were distributed into each well on the PCR array 96 well-plate. The qRT-PCR reaction proceeded according to a two-step cycling program on a Light Cycler 480 (*Roche*).

Light cycler program

temperature [°C]	time [min]	cycles	ramp rate [°C/sec]
95	10:00	1	1.0
95	00:15	45	1.0
60	01:00	1	1.0

For evaluation, the C_t values were exported into an Excel sheet and subjected to SABioscience's web-based PCR array data analysis tool (http://www.sabiosciences.com/pcr/arrayanalysis.php). Using this tool, fold changes in mRNA levels in B cells from *FasL ΔIntra* mice compared to wildtype ones were calculated based on the threshold cycle method.

2.2.7 Protein-biochemical methods

2.2.7.1 Protein-extraction from mammalian cells

For protein extraction, cells were harvested, washed with PBS and resuspended in lysis buffer containing SDS and protease inhibitors. Before use, 495 µl 10% SDS, 132 µl PMSF (100mM) and one tablet Complete Mini protease inhibitor cocktail were added to 9.5 ml lysis buffer (10x stock solution). The cells were disrupted by sonification (two times 15 seconds; 30% amplitude) and cellular debris was removed by centrifugation for ten minutes at 15,000 xg and 4°C.

2.2.7.2 Determination of protein concentration using the Bradford method

Protein concentration in the lysates was determined with a commerically available Bradford assay system (*BioRad*) that exploits the unspecific interaction of Coomasie blue dye with aromatic amino acids. A mixture containing 5 µl lysate, 200 µl Bradford reagent and 800 µl H_2O was incubated for five minutes at room temperature before being transferred into a plastic cuvette for spectrophotometric determination of the extinction at 595 nm. Calculation of protein concentration was based on a BSA standard curve.

2.2.7.3 SDS polyacrylamide gel electrophoresis (PAGE)

For the size-dependent electrophoretic separation of proteins, a defined amount of cell lysate was mixed with an equal volume of 2x sample buffer and boiled for five minutes at 95°C to denature the proteins. Samples were stored on ice until use. Stacking- and running-SDS-polyacrylamide gels were casted and allowed to polymerize for two hours, before 30-50 µl denatured sample were loaded and the gel was run at 80V until samples were concentrated at the end of the stacking gel (approximately 10-20 minutes) and then at 120V for sample separation. The Benchmark pre-stained protein ladder (*Invitrogen*) was used as a size standard.

2.2.7.4 Immunoblotting

Proteins separated by electrophoresis were transferred onto 0.45 µm nitrocellulose membranes using a semi-dry transfer system. The immunoblot was assembled between the electrodes of the transfer system by creating a stack of three buffer-soaked sheets of Whatmann paper, a piece of nitrocellulose membrane, the SDS-polyacrylamide gel and three final, buffer soaked sheets of Whatman paper. Proteins were transferred onto the membrane at 0.8 mA/cm^2 for two hours. Transfer efficiency and homologous loading of proteins were assessed by staining with Ponceau S.

For the immunoblot, membranes were incubated in blocking buffer for one hour at room temperature or overnight at 4°C to block unspecific binding. Antibodies against the proteins of interest were diluted in blocking buffer and incubated with the membrane overnight at 4°C. Following three washing steps, the membrane was incubated with an appropriate HRP-coupled antibody diluted in blocking buffer for one to two hours at room temperature. After three final washing steps, HRP-mediated chemiluminescence was detected with an ECL detection kit (*Amersham*).

2.2.8 *In vivo* studies

2.2.8.1 Analysis of thymocyte proliferation - Bromodesoxyuridine (BrdU) incorporation

To investigate the proliferation of thymocytes *in vivo*, incorporation of the thymidine analogue BrdU into newly synthesized DNA was analyzed. Three mice per group were intraperitoneally injected with 1 mg BrdU/100 µl PBS, followed by a second injection of 1 mg BrdU/100 µl PBS 30 minutes later. Control groups received twice 100 µl PBS instead of BrdU. The thymus was removed six hours after the second injection and was homogenized by grinding through gauze. After blocking Fc-receptors with FasFc (*BD Bioscience*) to prevent unspecific binding of antibodies, 10^6 total thymocytes were co-stained for CD4 and CD8 cell surface expression. BrdU incoporation into newly synthesized DNA was determined with the FITC BrdU Flow Kit according to the manufacturer's instructions (*BD Bioscience*). Cells stained for cell surface antigens were fixed and permeabilized with 100 µl Cytofix/Cytoperm buffer (20 minutes, room temperature). After washing with 1 ml Perm/Wash buffer, cells were incubated with 100 µl Cytoperm Plus buffer for ten minutes on ice. Following the treatment of cells with Cytoperm Plus buffer, cells need to be re-fixed. For this purpose, cells were washed, resuspended in 100 µl Cytofix/Cytoperm buffer, incubated for five minutes on ice and were washed again. To allow binding of the anti-BrdU antibody, the cellular DNA was digested with DNase I to expose the incorporated BrdU. This was accomplished by the addition of 30 µg DNase I (300 µg/ml) for one hour at 37°C. Samples were washed and stained with an anti-BrdU-FITC antibody (*BD Bioscience*) diluted 1:50 in 50 µl Perm/Wash buffer for 20 minutes at room temperature. Finally, cells were washed, resuspended in FACS buffer and analyzed on a FACSCalibur.

2.2.8.2 Expansion of Vβ8 T cell receptor chain-expressing T cells

Challenge with the bacterial superantigen *Staphylococcus* enterotoxin B (SEB) causes the specific expansion of Vβ8 T cell receptor chain-expressing T cells and can be used to assess the proliferative capacity of T cells *in vivo*. Three mice per group were intravenously injected with 100 µg SEB/200 µl PBS. Control groups received PBS only. Mice were sacrificed 48 hours post-injection and spleen and lymph nodes were removed. Single cell suspensions were prepared and stained for FACS analysis with fluorochrom-conjugated antibodies recognizing CD4, CD8 and Vβ8 or Vβ6. Staining for the Vβ6 TCR chain served as a control as Vβ6 TCR chain-expressing T cells do not respond to SEB challenge. For analysis of results, the increase in Vβ8 or Vβ6 TCR chain-expressing T cells in challenged *versus* unchallenged mice was calculated.

2.2.8.3 Germinal center formation

To compare the capacity of wildtype and *FasL ΔIntra* B cells to participate in thymus-dependent (TD) immune responses, mice were immunized and levels of responding B cell populations were determined by flow cytometry. For the immunization, 100 µg 4-hydroxy-3-nitrophenyl acetyl chicken gamma globulin (NP-CGG) per mouse were coupled to the adjuvant aluminium-potassium-dodecahydrate ($KAL(SO_4)_2$) by mixing equal volumes of NP-CGG (2 mg/ml) and 10% $KAL(SO_4)_2$ (in H_2O). After careful titration to pH 6.5 with 1N NaOH, the mixture was kept on ice for 30 minutes and was then centrifuged for ten minutes at 1,700 xg and 4°C. The pellet was washed three times with 1 ml PBS and resuspended in PBS to yield a concentration of 100 µg NP-CGG/200 µl PBS. Sterile filtrated, alum-precipitated NP-CGG was injected intrapertitoneally into five mice per group. Fourteen days post-immunization, mice were sacrificed, and 10^6 cells from spleen homogenates were stained for FACS analysis with 50 µl of the respective staining solution for 20 minutes at 4°C. Cell surface markers used to distinugish cell populations and set up of the specific staining reactions are described below. The dye TopRo-3 iodide was included in some stainings as a live/dead cell discriminator. All antibodies were purchased from *BD Bioscience* and had a concentration of 0.2 mg/ml.

Staining 1: B cells (B220$^+$), T cells (CD3$^+$) and B cell activation status (CD80$^+$, CD86$^+$)
Gating: for the discrimination of B and T cells: lymphocyte population (FSC/SSC dot plot); for B cell activation status: B220$^+$ lymphocytes

	dilution factor
CD86-FITC	1:300
CD3-PE	1:50
B220-PerCp	1:100
CD80-APC	1:1200

Staining 2: Germinal center B cells (PNA$^+$ CD95$^+$)
Gating: TopRo$^-$ B220$^+$ lymphocytes

	dilution factor
PNA-FITC	1:500
CD95-PE	1:75
B220-PerCp	1:100
TopRo-3 iodide	1:200

Staining 3: Plasma cells (Syndecan (CD138)$^+$, B220^{low})
Gating: TopRo$^-$ cells (TopRo/FSC dot plot)

	dilution factor
Syndecan (CD138)-Bio	1:250
SA-PE	1:400
B220-PerCp	1:100
TopRo-3 iodide	1:200

Staining 4: Follicular zone B cells (IgM^+ IgD^+) and marginal zone B cells (IgM^{high} IgD^{low})
Gating: Lymphocyte population

	dilution factor
IgD-FITC	1:800
IgM-PE	1:30
B220-PerCp	1:100
TopRo-3 iodide	1:200

Staining 5: Isotype class switch to IgG and antigen-specific B cells ($NP7^+$)
Gating: $B220^+$ lymphocytes

	dilution factor
IgG1-FITC	1:200
NP7-PE	1:400
B220-PerCp	1:100
IgM-APC	1:200

Additionally, antibody titers in serum of the immunized animals were determined. To obtain serum, heart blood was collected. Therefore, the heart was incised, blood was drawn with a Pasteur capillary and transferred into 1.5 ml reaction tubes. Blood was incubated on ice for at least 2-3 hours until the cruor had settled, and centrifuged for ten minutes at 15,000 xg (4°C). Supernatant was transferred into fresh 1.5 ml reaction tubes, cleared again and stored at -80°C until analysis.

To determine immunglobulin isotype concentrations and NP-specific antibodies solid phase sandwich ELISAs were performed. Incubations were done at room temperature and plates were washed between each assay step by addition of 200 µl PBS/well and subsequent decanting unless stated otherwise. Microtiter plates were coated with Ig-specific anti-mouse antibodies (anti-IgG1 or anti-IgM antibodies; c_{final} = 5 µg/ml) to determine antibody specificity, or, to determine antibody affinity, with NP3-BSA and NP17-BSA (c_{final} = 5 µg/ml) in carbonate buffer at 4°C overnight. Unspecific binding was blocked by incubation with 200 µl/well blocking buffer (one hour). Serum samples and standards (purified mouse IgM or IgG1; for the detection of NP-specific antibodies: standard serum) were pre-diluted in blocking buffer as follows:

	standard	serum sample
total IgG1 titer	1:100	1:50
total IgM titer	1:100	1:50
NP-specific IgG1 titer	1:1000	1:200
NP-specific IgM titer	1:200	1:50

One hundred microliters pre-diluted sample and standard were added to the top row on the microtiter plate and was serially diluted (1:2) in blocking buffer down to the bottom row using a 12-channel pipette, so that each well contained a total volume of 50 µl. Samples were allowed to bind for 1 hour at room temperature before the ELISA sandwich was formed by addition of 50 µl biotinylated secondary antibodies (anti-IgM-biotin or anti-IgG1-biotin, 1:500 in blocking buffer) for 30 minutes at room temperature. Following a 30 minutes incubation with 50 µl/well AP-coupled streptavidin (1:2000), 100 µl substrate solution (35 ml substrate-buffer/1 o-phenylendiamine tablet/21 µl H_2O_2) were added. AP-mediated substrate conversion was detected in an ELISA reader set to 405 nm. For evaluation, the OD readings were corrected for background signals. A standard curve was plotted with the OD readings represented on the y-axis and the concentration on the x-axis. Antibody concentration in the samples was calculated based on the linear regression of the standard curve.

2.2.8.4 Thioglycollate-induced peritonitis

Thioglycollate broth injection into the peritoneal cavity followed by peritoneal lavage represents a well-described model for the mimicry of a systemic bacterial infection. Innate host immune responses become activated and lead to an increase of local neutrophils within the first hours after infection.

Mice were injected intraperitoneally with 1 ml 3% thioglycollate broth in the evening of day one and treated a second time in the morning of the following day. Control groups were left untreated. Three hours after the second injection, mice were sacrificed. The animals were carefully skinned to keep the peritoneum intact. Incissions were made at both flanks of the mouse and across the abdomen, so that the skin could be detached from the peritoneum and be peeled off towards the head. For peritoneal lavage, 10 ml sterile Hank's balanced salt solution (HBSS)/0.25% BSA were injected into the peritoneal cavity. Therefore, the syringe was pierced through the abdominal fat, since the fat serves as a natural plug preventing leakage of administered fluid. Cells were washed out by massaging the peritoneum for two minutes. Finally, the cell-containing fluid was recollected from the peritoneal cavity using a syringe. The number of infiltrating neutrophils was determined as the percentage of Ly6G (Gr-1)-expressing cells by FACS analysis of 10^6 cells.

2.2.8.5 Statistics
Statistical analysis was performed using two-tailed Student's t-test.

3 Results

3.1 Mouse model for FasL reverse signaling deficiency

To study the physiological consequences of FasL reverse signaling *in vivo*, a mouse model for FasL reverse signaling deficiency was established in the group of PD Dr. Martin Zörnig (Georg-Speyer-Haus, Frankfurt). In these mice, the *FasL* gene was replaced by a truncated version that completely lacks the ICD (Δ aa 1-74) while it still expresses the wildtype transmembrane and extracellular domain (**Fig. 3.1**).

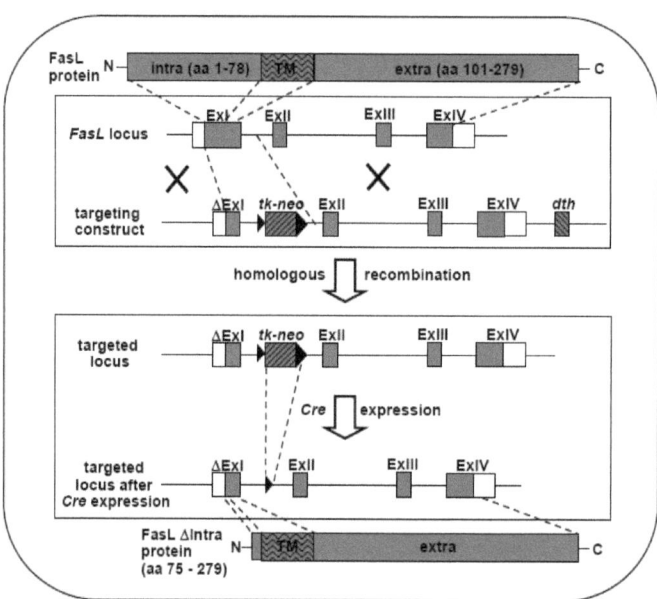

Figure 3.1 Mouse model for FasL reverse signaling deficiency.
The *FasL ΔIntra* mouse line was established by replacing the wildtype FasL with a truncated version lacking the FasL ICD (Δ aa 1-74). Homologous recombination of the targeting construct with the endogenous *FasL* locus deleted the first 74 N-terminal amino acids in exon 1 and introduced a floxed *tk-neo* cassette for positive selection into intron 1. Chimeric mice carrying the truncated *FasL ΔIntra* allele were crossed with the early deleter transgenic mouse strain *CMV-Cre* to remove the *tk-neo* cassette and subsequently were bred into the C57Bl/6N genetic background (at least eight generations). Intra = intracellular domain; TM = transmembrane domain; extra = extracellular domain; Ex = exon; *tk-neo* = Thymidine kinase-neomycine resistance cassette (positive selection); *dth* = *Diphteria toxin A* gene cassette (negative selection); solid black arrowheads = *loxP* sequences

Heterozygous mice ($FasL^{Wt/\Delta}$) were crossed, and resulting offspring were genotyped to identify wildtype and homozygous $FasL$ $\Delta Intra$ chimeras. PCR with genomic DNA isolated from tail biopsies revealed that all offspring were viable. Inheritance followed a clear Mendelian pattern, i.e. an inheritance ratio of wildtype to heterozygous to homozygous mutant mice of 1:2:1. Of 1200 animals, 278 mice carried two wildtype alleles ($FasL^{Wt/Wt}$) and 293 possessed two $FasL$ $\Delta Intra$ alleles ($FasL^{\Delta/\Delta}$), while 625 mice were heterozygous ($FasL^{Wt/\Delta}$). This corresponds to a ratio of the different genotypes of 1:2.3 :1.1 (wildtype:heterozygous:homozygous; **Fig. 3.2**).

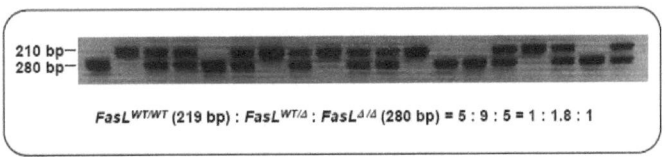

$FasL^{WT/WT}$ (219 bp) : $FasL^{WT/\Delta}$: $FasL^{\Delta/\Delta}$ (280 bp) = 5 : 9 : 5 = 1 : 1.8 : 1

Figure 3.2 Genotyping and Mendelian inheritance pattern of $FasL$ $\Delta Intra$ mice. Homozygous mice ($FasL^{Wt/\Delta}$) were crossed, and offspring were routinely genotyped by PCR using genomic DNA as template and primers which amplify the region around the remaining loxP sequence in intron 1 of the mutated allele. As only the targeted $FasL$ locus harbours the loxP site, the resulting PCR product is elongated ($FasL^{\Delta/\Delta}$: 280 bp vs. $FasL^{Wt/Wt}$: 219 bp). A representative picture of an agarose-gel electrophoresis of PCR products is shown. Of 19 animals analyzed, five mice carry two wildtype alleles ($FasL^{Wt/Wt}$) and five possess two $FasL$ $\Delta Intra$ alleles ($FasL^{\Delta/\Delta}$), while nine mice are heterozygous ($FasL^{Wt/\Delta}$).

3.2 $FasL$ $\Delta Intra$ mice represent a suitable model to study FasL reverse signaling

3.2.1 $FasL$ $\Delta Intra$ mice express functional FasL that is capabale of inducing apoptosis

To verify that the $FasL$ $\Delta Intra$ mouse line represents a suitable model to study FasL reverse signaling, FasL expression was investigated. This is an important point, as $FasL^{-/-}$ mice and Fas or FasL-defective mice (the natural mouse mutants Fas^{lpr}, $FasL^{gld}$ and the genetic Fas and $FasL$ knockout models) develop a severe lymphoproliferative disease with the occurence of an aberrant T cell population, auto-antibodies and splenomegaly (Nagata et al., 1995). The lethal pathology is mostly attributed to the lack of Fas-mediated apoptosis. By RT-PCR analysis using primers that bind sequences in the 5'-UTR (screen-for) and near the 3'-end of exon 1 (screen-rev), the presence of wildtype and truncated FasL transcript in activated B and T cells could be proven (**Fig. 3.3 A**). Comparable expression of FasL protein at the cell surface of activated splenic lymphocytes became evident by flow cytometric staining using the anti-mFasL antibody clone MFL3 (**Fig. 3.3 B**).

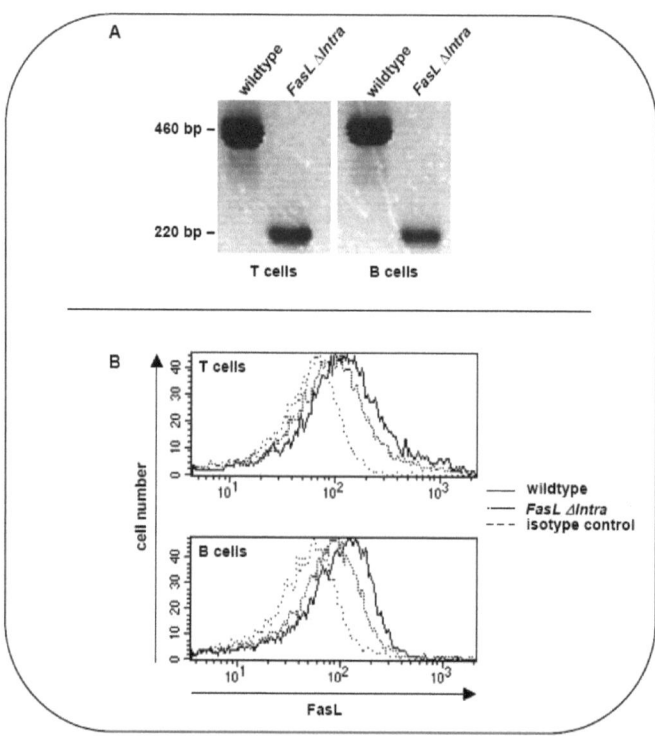

Figure 3.3 FasL expression in *FasL ΔIntra* mice.
(**A**) *FasL* mRNA expression from the wildtype or targeted *FasL* locus was analyzed by RT-PCR. Using total RNA isolated from activated T and B cells (1 µg/ml anti-CD3 or anti-IgM antibody, for four hours) as template and primers that bind sequences in the 5'-UTR and near the 3'-end of exon 1 resulted in a 460 bp fragment for the wildtype and a 220 bp fragment for the *FasL ΔIntra* transcript. (**B**) Expression of FasL at the cell surface of activated lymphocytes. Lymphocytes were activated by stimulation with 5 µg/ml anti-CD3 antibody (T cells) or 10 µg/ml LPS (B cells) for 24 hours before staining with PE-conjugated anti-FasL (MFL-3) or isotype control antibody. One representative experiment out of four is shown. Wildtype cells are depicted by the solid, those of *FasL ΔIntra* mice by the dotted, and the isotype control by the wide dashed line.

Co-culture assays employing re-stimulated T cell blasts as effectors and Fas-sensitive target cells (B cell lymphoma cell line A20) indicated that the expressed *FasL ΔIntra* protein is functional in terms of its apoptosis-inducing function (**Fig. 3.4**). Notably, the killing capacity of *FasL ΔIntra*-effectors is reduced by approximately 40% compared to that of wildtype cells, but still is significantly higher than that of homoyzgous and heterozygous $FasL^{gld}$ mice. The latter ($FasL^{Wt/gld}$) are rescued from development of the lymphoproliferative disease observed in homozygous animals by the presence of the single wildtype allele. Thus, the expression of a functional FasL (in terms of its apoptosis-inducing function) at the cell surface of activated lymphocytes makes the newly established *FasL ΔIntra* mouse line a suitable model to study the physiological consequences of FasL reverse signaling *in vivo*.

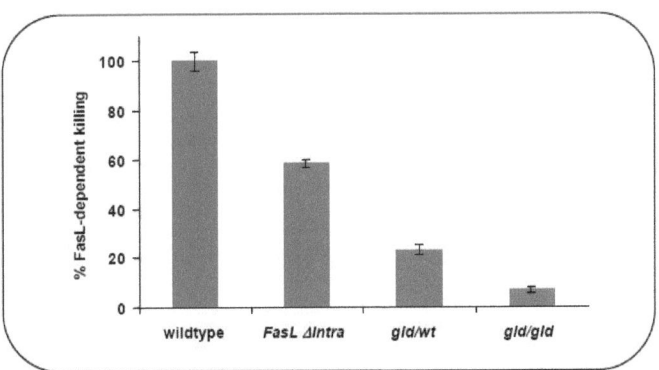

Figure 3.4 *FasL ΔIntra* **mice express a functional FasL protein capable of inducing apoptosis.** The capacity of FasL-dependent killing was evaluated in co-culture experiments using re-stimulated (10 µg/ml anti-CD3 antibody for 24 hours) T cell blasts of wildtype, *FasL ΔIntra* and $FasL^{gld}$ mice as effectors and Fas-sensitive, CFSE-labeled target cells (A20) in an effector : target ratio of 2.5 : 1. After 6 hours, the extent of target cell death was assessed by PI staining of ethanol-fixed cells. Killing capacity is expressed as the percentage of cells with hypoploid DNA content (sub-G1 cells) minus the percentage of sub-G1 cells when FasL-dependent killing was blocked by addition of 100 µg/ml FasFc. Columns represent the mean values and error bars the standard error of the mean (SEM) of four individual experiments.

3.2.2 *FasL ΔIntra* mice do not display obvious phenotypic anomalies

In a broad, standardized phenotypic screen performed at the German Mouse Clinic (GMC, Munich) with five to 21 week old, unchallenged mice, no relevant differences between *FasL ΔIntra* and wildtype mice could be detected (data not shown). Likewise, an extensive analysis of main T and B cell compartments in spleen, lymph nodes, peritoneal cavity, blood and bone marrow of 8-12 week old mice revealed comparable levels of each population inspected (**Table 3.1**).

Table 3.1 Unchallenged *FasL ΔIntra* mice do not show any phenotypic anomalies.
B and T cell populations in spleen, lymph nodes, bone marrow and peritoneal cavity were analyzed by flow cytometry by staining for characteristic cell surface markers as indicated. Mice were used at an age of 8-12 weeks. Numbers represent the mean values and SEM of at least four mice per group.

Spleen

		Marker profile	wildtype	*FasL ΔIntra*
Lymphocyte populations	T cells	$CD3^+$	29.54 ± 3.31	32.12 ± 0.88
	$CD4^+$ T cells	$CD3^+ CD4^+$	63.04 ± 4.60	61.64 ± 1.67
	$CD8^+$ T cells	$CD3^+ CD8^+$	30.87 ± 4.10	31.02 ± 1.59
	B cells	$B220^+$	53.82 ± 5.02	59.19 ± 1.23
Activation status	T cells	$CD4^+$ or $CD8^+$, $CD69^+$	7.50 ± 1.50	7.52 ± 1.29
	B cells	$B220^+CD86^+$	10.36 ± 0.13	11.02 ± 0.26
Fas expression	T cells	$CD4^+$ or $CD8^+$, $CD95^+$	26.00 ± 3.00	27.00 ± 2.00
	B cells	$B220^+CD95^+$	22.99 ± 2.69	23.87 ± 2.21
Lymphocyte subsets	$CD4^+$ T cells, naïve	$CD4^+CD62L^+CD44^-$	70.59 ± 2.69	67.30 ± 4.08
	$CD4^+$ T cells, central memory	$CD4^+CD62L^+CD44^+$	3.03 ± 1.63	4.74 ± 0.69
	$CD4^+$ T cells, effector memory	$CD4^+CD62L^-CD44^+$	16.12 ± 5.65	22.40 ± 2.99
	$CD8^+$ T cells, naïve	$CD8^+CD62L^+CD44^-$	59.04 ± 5.04	51.23 ± 4.74
	$CD8^+$ T cells, central memory	$CD8^+CD62L^+CD44^+$	10.97 ± 5.91	15.90 ± 2.51
	$CD8^+$ T cells, effector memory*	$CD8^+CD62L^-CD44^+$	3.61 ± 1.07	8.71 ± 1.58
	Transitional B cells T1	$AA1.4^+CD23^-IgM^{high}$	22.74 ± 1.59	22.05 ± 1.78
	Transitional B cells T2	$AA1.4^+CD23^+IgM^{high}$	21.78 ± 0.86	24.13 ± 1.24
	Transitional B cells T3	$AA1.4^+CD23^+IgM^{low}$	17.69 ± 1.52	19.19 ± 1.57
	Marginal zone B cells	$B220^+IgM^{high}IgD^{low}$	7.31 ± 0.90	8.92 ± 1.15
	Follicular B cells	$B220^+IgM^+IgD^+$	80.65 ± 0.18	80.27 ± 0.88

* The difference in effector memory $CD8^+$ T cells between *FasL ΔIntra* and wildtype mice is significant (p = 0.5).

Other lymphoid organs

		Marker profile	wildtype	FasL ΔIntra
Lymph node	T cells	CD3$^+$	66.19 ± 2.55	70.52 ± 2.57
	total CD4$^+$ T cells	CD3$^+$CD4$^+$	58.08 ± 3.09	57.43 ± 2.95
	total CD8$^+$ T cells	CD3$^+$CD8$^+$	38.84 ± 2.75	39.07 ± 3.35
	B cells	B220$^+$	28.52 ± 0.30	26.36 ± 1.82
	Follicular B cells	B220$^+$IgM$^+$IgD$^+$	27.31 ± 0.70	25.06 ± 1.27
	Activated B cells	B220$^+$CD86$^+$	9.77 ± 0.31	10.17 ± 0.26
	Fas expression	B220$^+$CD95$^+$	20.67 ± 1.33	19.49 ± 0.44
Bone marrow	Pro B cells	B220^{low}CD43^{low}	3.83 ± 0.05	4.77 ± 0.78
	Pre Bcells	B220^{low}CD43$^-$	32.83 ± 3.55	25.90 ± 3.52
	Immature B cells	B220$^+$IgM$^+$IgD$^-$	16.19 ± 0.27	11.46 ± 3.44
	Recirculating B cells	B220$^+$IgM$^+$IgD$^+$	32.64 ± 2.96	35.49 ± 3.37
Peritoneal cavity	T cells	CD3$^+$	6.11 ± 0.45	6.23 ± 0.20
	B1a B cells	CD5$^+$B220^{low}	32.07 ± 4.47	33.42 ± 4.13
	B1b B cells	CD5$^-$B220^{low}	7.96 ± 0.66	6.38 ± 1.05
	B2 B cells	CD5$^-$B220$^+$	40.12 ± 4.37	40.13 ± 5.38

The life expectancy of mutant mice could not be determined, as mice were sacrificed before death of natural causes. However, even in old animals (12-18 month) no relevant difference regarding phenotype and viability was found between *FasL ΔIntra* and wildtype mice. Importantly, the old *FasL ΔIntra* mice did not develop any signs of the lymphoproliferative disease associated with the *FasL$^{gld/gld}$*- and *Fas$^{lpr/lpr}$*-mutation, as indicated by the absence of anti-dsDNA auto-antibodies, rheumatoid factor, CD3$^+$B220$^+$CD4$^-$CD8$^-$ T cells or splenomegaly (**Table 3.2**). The absence of any obvious phenotypic anomalies supports the notion that the *FasL ΔIntra* mouse line is suitable to decipher the physiological function of the FasL ICD.

Table 3.2 FasL ΔIntra mice do not develop signs of the $FasL^{gld/gld}$-phenotype.
Enlargement of the spleen was assessed by visual inspection. Of note, $FasL^{gld/gld}$ mice exhibited splenomegaly, while three FasL ΔIntra mice had only slightly enlarged spleens. Cell viability and organ cell counts were measured with a Casy cell counter. Aberrant T cells were detected by flow cytometric staining of splenocytes and thymocytes for cell surface markers as indicated. All mice were used at an age of 12-18 month. Mean values and SEM of at least three mice per group are shown. S=spleen; T=thymus; sn=statistically significant difference in wildtype or FasL ΔIntra mice vs. $FasL^{gld/gld}$ mice; sn*=statistically significant in wt vs. $FasL^{gld/gld}$; n.d.=not determined

	organ	Marker profile	wildtype	FasL ΔIntra	$FasL^{gld/gld}$	
Spleen enlarged			0/4	3/16	4/4	
Viability	S		93.53 ± 0.83	93.16 ± 3.84	88.36 ± 1.28	sn
Cell counts [*10⁷]	S		20.82 ± 2.47	25.80 ± 8.38	53.50 ± 16,27	sn*
	T		3.48 ± 0.76	5.87 ± 1.10	19.33 ± 09.53	sn
Aberrant T cell population	S	$CD3^+B220^+$	2.00 ± 0.41	2.50 ± 0.34	23.25 ± 06.02	sn
	T	$CD3^+B220^+$	0.43 ± 0.07	n.d.	41.00 ± 02.45	sn
		$Thy1^+CD4^-CD8^-$	n.d.	4.00 ± 0.32	22.67 ± 10.20	sn
		$Thy1^+CD4^+CD8^+$	n.d.	83.40 ± 1.03	3.67 ± 02.67	sn

3.3 Analysis of FasL reverse signaling *ex vivo*

3.3.1 The FasL ICD impairs lymphocyte proliferation

Several reports have implicated FasL reverse signaling in the regulation of T cell proliferation and activation (Lettau et al., 2009). Based on these findings, the proliferation of different lymphocyte populations was investigated in FasL ΔIntra mice. Primary lymphocytes were isolated from the spleens of wildtype and mutant mice using magnetic bead purification and lineage specific kits for the negative selection of the desired cell subset. A purity of lymphocyte preparations of over 90% was usually achieved (data not shown). CFSE-labeling experiments revealed that the proliferation of B cells, $CD4^+$ T cells and $CD8^+$ T cells in response to mitogenic stimulation for 72 hours was strongly enhanced in the absence of the FasL ICD (**Fig. 3.5 A**). This effect was most pronounced in B cells. Interestingly, differences in the proliferative capacity of $CD4^+$ T cells could only be detected after the depletion of $CD25^+CD4^+$ regulatory T cells. Thus, the presence of the FasL ICD appears to negatively impact lymphocyte proliferation. In accordance with these results, cells completely lacking FasL (*FasL*[-/-]) (Karray et al., 2004; Mabrouk et al., 2008) responded even stronger to mitogenic stimulation than those lacking only the FasL ICD (**Fig. 3.5 B**).

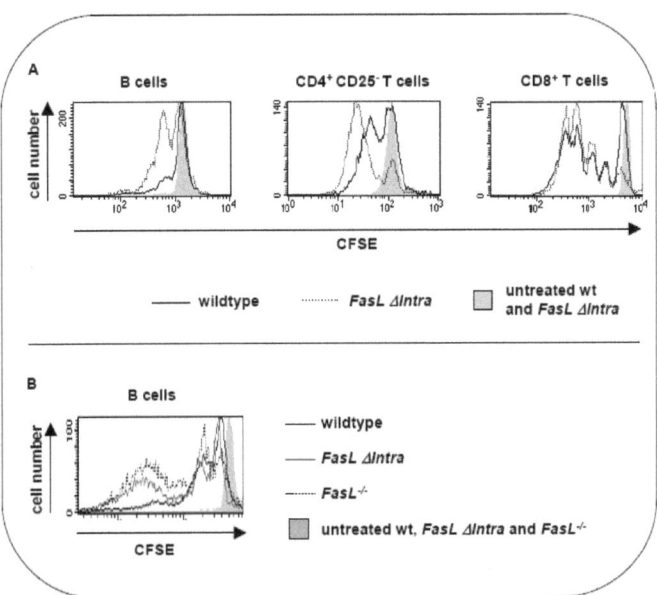

Figure 3.5 Increased proliferation in the absence of the FasL ICD.
CFSE-labeled lymphocytes were left untreated or were stimulated with 5 µg/ml anti-CD3 antibody (T cells) or 10 µg/ml anti-CD40 antibody (B cells) for 72 hours. Proliferation of living cells was measured by FACS analysis as the decrease in fluorescence intensity of the CFSE-staining. The grey peak represents non-stimulated lymphocytes that did not proliferate. **(A)** In the absence of the FasL ICD (dotted line), lymphocyte proliferation is enhanced compared to wildtype cells (solid line). One representative experiment out of five is shown. **(B)** B cells completely lacking FasL (*FasL*[-/-]) responded even stronger to mitogenic stimulation than those lacking only the FasL ICD. One experiment out of two is shown.

3.3.2 Defective FasL reverse signaling accounts for the observed enhanced proliferation of *FasL ΔIntra* lymphocytes

3.3.2.1 Comparable extent of cell death in *FasL ΔIntra* and wildtype lymphocytes

In the immune system, activation-induced cell death (AICD), *i.e.* deletion of re-stimulated lymphocytes, plays a crucial role in the regulation of clonal expansion/deletion of effector cells and the termination of immune responses. AICD proceeds through death receptor-mediated apoptosis, and the FasL/Fas system plays a pivotal role in its course (Green *et al.*, 2003; Krammer *et al.*, 2007). Therefore, what appears as an enhanced proliferation might alternatively reflect a survival advantage instead.

To eliminate this possibility, cell death in response to TCR or BCR cross-linking was analyzed by Annexin V/PI staining. The extent of cell death in T and B cells in response to stimulation for zero, 48 or 72 hours was not affected by the absence or presence of the FasL ICD (**Fig. 3.6 A**). Likewise, the number of viable T cell blasts upon re-stimulation (1 μg/ml anti-CD3 antibody for 24 hours) was decreased by approximately 30% in both, mutant and control cells (**Fig. 3.6 B**). Therefore, the enhanced proliferation in *FasL ΔIntra* lymphocytes does not reflect decreased cell death.

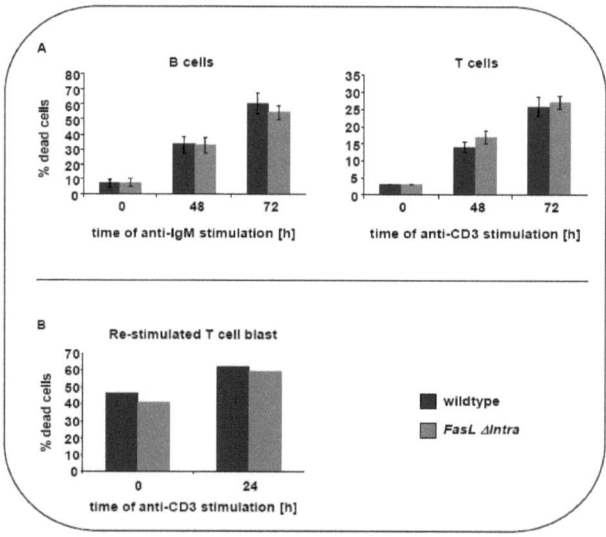

Figure 3.6 Comparable extent of cell death in lymphocytes of *FasL ΔIntra* and wildtype mice. Cell death was quantified by staining with Annexin-V/PI and FACS analysis. Diagrams display the percentage of positively stained dead cells as the percentage of dead cells. **(A)** Naïve lymphocytes were stimulated with 1 µg/ml of either anti-IgM antibody (B cells) or anti-CD3 antibody (T cells) for the indicated times. Columns represent mean values and bars represent the SEM of five individual experiments. **(B)** AICD in T cell blasts re-stimulated with 1 µg/ml anti-CD3 antibody for 24 hours. One single experiment was performed.

3.3.2.2 Non-apoptotic signaling through Fas does not account for the proliferative differences in *FasL ΔIntra* and wildtype lymphocytes

To exclude signaling through the Fas receptor (instead of FasL reverse signaling) as a cause for the differential response, activation-induced proliferation was investigated in a Fas signaling-defective setting. Crossing *FasL ΔIntra* with *Fas^{lpr}* mice resulted in offspring homozygously harboring the mutant *Fas^{lpr}* and either the wildtype or the truncated *FasL* locus (*Fas$^{lpr/lpr}$/FasL$^{Wt/Wt}$* and *Fas$^{lpr/lpr}$/FasL$^{Δ/Δ}$*). Using the same experimental set up as in previous experiments, it was demonstrated that the proliferative differences between mutant and wildtype lymphocytes could be reproduced in the absence of a functional Fas receptor (**Fig. 3.7 A**), pointing to a causative role for FasL reverse signaling. In support of this notion, levels of Fas at the cell surface of resting and activated lymphocytes are comparable between *FasL ΔIntra* (*Fas$^{Wt/Wt}$*) and wildtype (*Fas$^{Wt/Wt}$*) mice (**Fig. 3.7 B**).

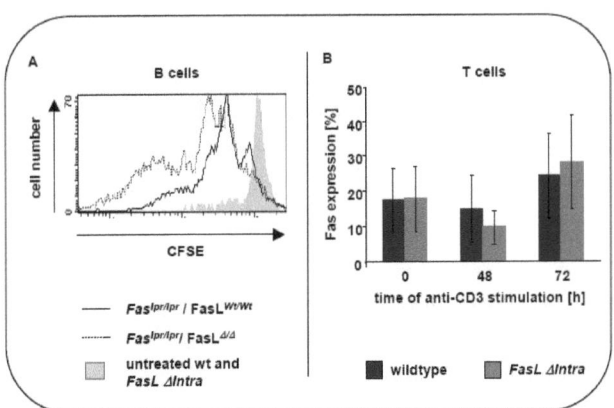

Figure 3.7 Signaling through Fas does not account for proliferative differences in *FasL ΔIntra* and wildtype lymphocytes. (A) B cell proliferation in a Fas signaling-deficient model. *FasL ΔIntra* mice were mated with *Fas$^{lpr/lpr}$* animals to obtain homozygous *Fas$^{lpr/lpr}$* offspring that express either wildtype or truncated FasL (*Fas$^{lpr/lpr}$/FasL$^{Wt/Wt}$, Fas$^{lpr/lpr}$/FasL$^{Δ/Δ}$*). Activation-induced proliferation of B cells (10 μg/ml anti-CD40 antibody, 72 hours) was investigated in CFSE-labeling experiments. Resting lymphocytes are represented by the filled grey curve. The solid black line represents B cells with wildtype FasL, the dotted line those prepared from *FasL ΔIntra* mice. One representative experiment out of three is shown. **(B)** Comparable levels of Fas at the cell surface of T cells. Splenic T cells were isolated and stimulated for the indicated times with 1 μg/ml anti-CD3 antibody. The percentage of Fas-expressing T cells was analyzed by flow cytometry using an anti-Fas (JO2) antibody. Columns represent mean values and bars represent the SEM of three independent experiments.

3.3.3 The FasL ICD regulates ERK1/2 activation and proliferation by influencing phosphorylation of PLCγ2 and PKC

Upon antigen encounter, lymphocytes acquire an activated phenotype, characterized by a blastic morphology and the expression of activation markers at the cell surface, and they expand clonally. At the molecular level, these changes correlate with the activation of various signaling pathways, such as the Mitogen-activated protein kinase (MAPK)-, the Phosphoinositide-specific phospholipase C (PLC)/Protein kinase C (PKC)- and the Phosphoinositide 3-kinase (PI3K)/Akt-pathway.

3.3.3.1 The presence of the FasL ICD reduces ERK1/2 activation

To identify a molecular correlate for the differential proliferative capacity of *FasL ΔIntra* and wildtype lymphocytes, activation-induced ERK1/2 phosphorylation was analyzed as a prototypic marker for cell proliferation (Krishna et al., 2008). In cell-based ELISAs, phosphorylation of ERK1/2 (T202/Y204) in B cells upon BCR cross-linking with anti-IgM antibody for various time points was quantified. As expected, mitogenic stimulation resulted in a strong increase of phosphorylated ERK1/2. While initial levels of phosphorylated protein were comparable, mitogenic stimulation led to significantly higher levels of activated ERK1/2 in *FasL ΔIntra* B cells compared to wildtype cells (**Fig. 3.8 A**). This finding was confirmed by Western blot experiments in which ERK1/2 phosphorylation in lysates from activated B cells was detected with an anti-pERK1/2 antibody (**Fig. 3.8 B**). Both experiments revealed that in the absence of the FasL ICD, ERK1/2 activation was already elevated five minutes after stimulation and remained elevated at all time points analyzed, as compared to the presence of the full length protein in wildtype cells.

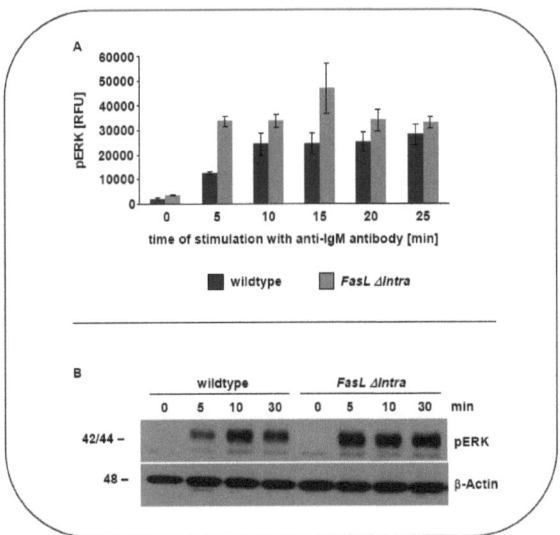

Figure 3.8 The presence of the FasL ICD reduces ERK1/2 activation.
(A) B cells were stimulated with 1 µg/ml anti-IgM antibody for the indicated time points, and ERK1/2 phosphorylation was analyzed by ELISA. The diagram displays the relative induction of ERK1/2 activation, measured as the increase in relative fluorescence intensity of phosphorylated ERK1/2 compared to total ERK1/2 protein. Data are presented as the mean values and SEM of four independent experiments. (B) Western blot analysis of ERK1/2 phosphorylation in lysates of B cells stimulated with 1 µg/ml anti-IgM antibody for the indicated times. Phosphorylated ERK1/2 was detected with an anti-pERK1/2 antibody. β-Actin served as loading control. One representative experiment out of two is shown.

3.3.3.2 MAP-kinases upstream of ERK1/2 do not appear to be regulated by FasL reverse signaling

Further characterizing the MAPK-pathway upstream of ERK1/2, the protein kinases c-Raf and MEK1/2 were analyzed in analogous experimental set ups. Although mitogenic stimulation again induced phosphorylation (c-Raf: S38; MEK1/2: S218/S222), neither c-Raf nor MEK1/2 were differentially regulated in B cells isolated from *FasL ΔIntra* mice compared to those from wildtype mice (**Fig. 3.9 A**). When activation-induced proliferation was investigated in the presence of the MEK1/2 inhibitor U0126, it could be shown that B cell proliferation was markedly impaired, but not completely abolished (**Fig. 3.9 B**). These findings suggested, that B cell proliferation is partially MAPK-dependent, but that the MAPK-cascade upstream of ERK is not subject to FasL reverse signaling-mediated regulation.

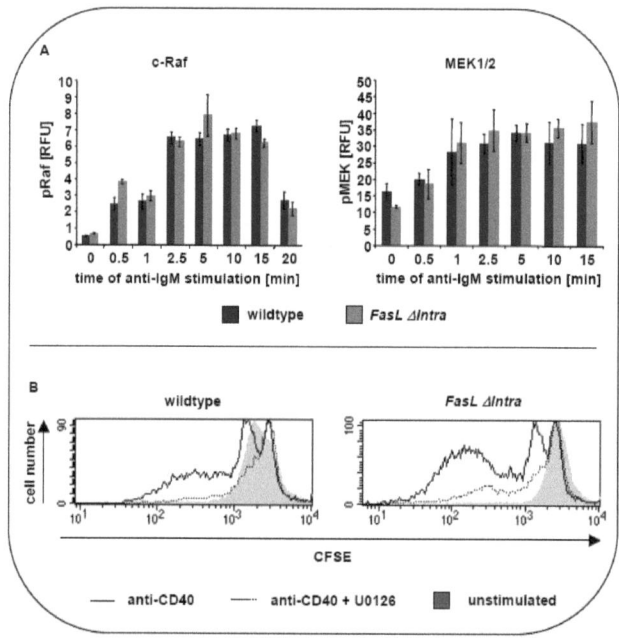

Figure 3.9 Neither c-Raf nor MEK1/2 are regulated by FasL reverse signaling.
(**A**) Splenic B cells were stimulated with 1 µg/ml anti-IgM antibody for the indicated time points and analyzed for c-Raf (left) or MEK1/2 (right) phosphorylation by ELISA. The diagrams display the relative induction of specific protein activation, measured as the ratio of the relative fluorescence intensities of phospho-protein *vs*. total protein. Mean values and SEM of eight (c-Raf) or four (MEK1/2) experiments are shown. (**B**) B cell proliferation is partially MAPK-dependent. Activation-induced (10 µg/ml anti-CD40 antibody, 72 hours) B cell

proliferation in the absence (solid line) or presence (dashed line) of the MEK inhibitor U0126 (10µM) was asessed in CFSE-labeling experiments. One experiment out of two is shown.

3.3.3.3 FasL reverse signaling via PLCγ2 and PKC regulates ERK1/2 activation and proliferation

In agreement with a partial dependence on the MAPK pathway, PMA/ionomycine-mediated proliferation of B cells was not affected by the MEK1/2 inhibitor U0126 (**Fig. 3.10**). As stimulation with PMA/ionomycine specifically triggers the activation of PKC (Kurosaki et al., 2000; Paulsen et al., 2009), this finding indicates an involvement of the PLC/PKC pathway in the regulation of BCR-induced proliferation of wildtype and FasL ΔIntra B cells.

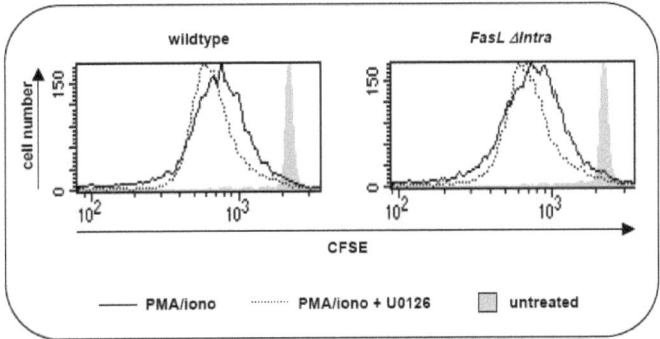

Figure 3.10 PMA/ionomycine-mediated proliferation is not affected by MEK inhibition. Splenic B cells were stimulated with 100 ng/ml PMA and 0.2 µg/ml ionomycine for 72 hours in the absence (solid line) or presence (dashed line) of the MEK1/2 inhibitor U0126 (10µM). Proliferation was quantified in CFSE-dilution assays by FACS analysis. One experiment of two is shown.

Indeed, intracellular stainings for phosphorylated PLCγ2 (Y759) and PKC (T497) after BCR cross-linking revealed an enhanced phosphorylation of both molecules in FasL ΔIntra B cells compared to wildtype cells (**Fig. 3.11 A**). To link PKC activity to ERK phosphorylation and proliferation, the effects of the general PKC inhibitor GF 109203x hydrochloride were determined (**Fig. 3.11 B**). PKC inhibition severely impaired ERK activation as well as proliferation of B cells derived from *FasL ΔIntra* and from wildtype mice. These results suggest an important role for PKC activity in FasL reverse signaling.

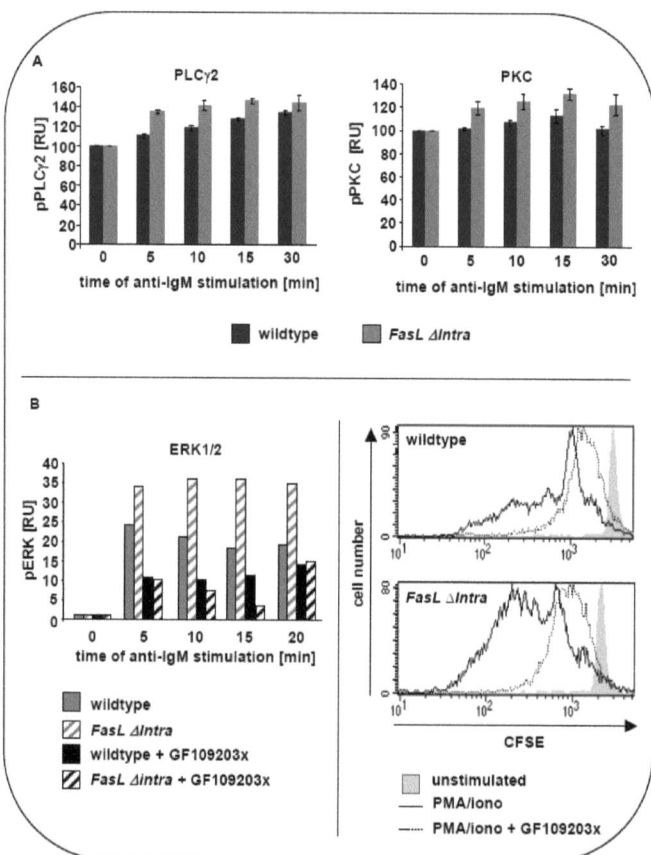

Figure 3.11 PLCγ2/PKC signaling regulates ERK1/2 activation and proliferation in B cells. (A) The activation of PLCγ2 and PKC is repressed by FasL reverse signaling. B cells were stimulated with 1 µg/ml anti-IgM antibody for the indicated times. Activation of PLCγ2 (left) and PKC (right) was evaluated by intracellular staining with anti-phospho- PLCγ2 and anti-phospho-PKC antibodies and subsequent FACS analysis. The geometric mean fluorescence intensity (MFI) was related to the MFI of untreated cells. Data are expressed as the mean values and SEM of four independent experiments. **(B)** *left:* PKC inhibition impairs ERK1/2 activation. ERK1/2 activation in response to B cell stimulation with 1 µg/ml anti-IgM antibody in the presence (dashed bars) or absence (filled bars) of the PKC inhibitor GF109203x hydrochloride (1µM) was determined by ELISA. The diagram shows the relative

induction of ERK1/2 activation, measured as the ratio of relative fluorescence intensities of phosphorylated ERK1/2 to total ERK1/2 protein. One experiment out of four is shown. *right*: PKC inhibition impairs B cell proliferation. PKC-dependent B cell proliferation (100 ng/ml PMA and 0.2 µg/ml ionomycine, 72 hours) in the presence (dotted line) or absence (solid line) of 1µM GF109203x hydrochloride was measured in CFSE-dilution assays. Unstimulated cells are represented by the grey curve. One representative experiment out of four is shown.

3.3.3.4 The PI3K/Akt pathway is apparently not involved in FasL reverse signaling

To explore a potential involvement of a third pathway important for lymphocyte activation in FasL ICD-mediated signaling, signal transduction *via* the PI3K/Akt pathway was analyzed. Using phosphorylation of the protein kinase Akt (S473) as readout in cell-based ELISA experiments, it could be demonstrated that kinetics and extent of Akt activation in B cells in response to mitogenic stimulation were not affected by the presence or absence of the FasL ICD (**Fig. 3.12**).

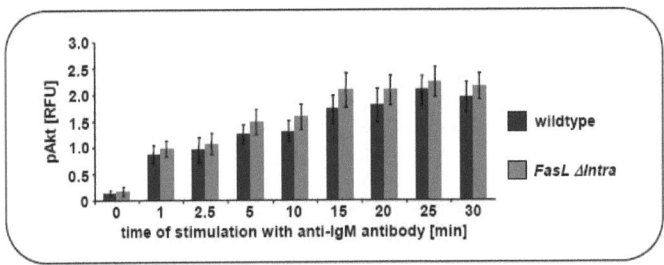

Figure 3.12 The PI3K/Akt pathway is not involved in FasL reverse signaling.
B cells were stimulated with 1 µg/ml anti-IgM antibody for the indicated time points, and Akt phosphorylation was analyzed by ELISA. The diagram shows the relative induction of Akt activation measured as the ratio of relative fluorescence intensities of phosphorylated Akt *vs.* total Akt protein. Data are expressed as the mean values and SEM of eight independent experiments.

3.3.4 Identification of FasL reverse signaling target genes

Previous studies in our group revealed a Notch-like proteolytic processing of FasL (Kirkin et al., 2007) and identified the transcription factor Lef-1 as an interaction partner of the FasL ICD (Lückerath et al., manuscript submitted). Additionally, data presented in this study show a FasL ICD-dependent differential activation of the transcriptional regulator ERK1/2. These findings promted the search for genes that are targeted by FasL reverse signaling.

3.3.4.1 Global gene expression profiling

Molecular phenotyping was performed to screen for potential target genes on a close to genome-wide basis. Co-operating with the German Mouse Clinic, cDNA derived from activated or resting T cells of either *FasL ΔIntra* or wildtype mice was hybridized to Affymetrix 21K cDNA chips (21,000 genes covered). In resting cells, the expression of only 22 genes deviated slightly, with 0.4-0.7 fold changes in relative transcript levels (appendix **Table A.1**). In contrast, 181 genes were found to be significantly regulated (up to eight fold) in activated lymphocytes (appendix **Table A.2**). None of the genes present on the array was regulated in both, unstimulated and stimulated, samples, indicating specific signals. Despite the large number of genes identified, cluster analysis did not reveal the regulation of any particular functional gene group.

3.3.4.2 FasL reverse signaling regulates genes associated with lymphocyte proliferation and activation

To identify *bona fide* target genes of FasL ICD-mediated singaling, the differential expression of candidate genes was verified by quantitative real-time PCR (qRT-PCR). Candidates were chosen among the genes which differed mostly between wildtype and *FasL ΔIntra* mice in microarray screens. By default, relative expression levels were normalized to the housekeeping gene *Hypoxanthine phosphoribosyltransferase 1* (*Hprt1*) and calculated using the threshold cycle method. Interestingly, when comparing activated B cells from *FasL ΔIntra* mice with those form wildtype mice, various genes associated with lymphocyte proliferation and activation were significantly (≥ two-fold) up-regulated in *FasL ΔIntra* B cells (**Fig 3.13 A**), for example *Nuclear factor kappa-light-chain-enhancer of activated B cells* (*NF-κb1*), *Nuclear factor of activated T cells 1* (*NFAT1*) and *Interferon regulatory factor 4* (*Irf4*), providing an explanation for the enhanced proliferative response in the absence of the FasL ICD. Importantly, results could be reproduced when gene expression was normlized to an alternate housekeeping gene (G*lycerinaldehyde-3-phosphate-dehydrogenase; Gapdh*) (**Fig. 3.13 B**).

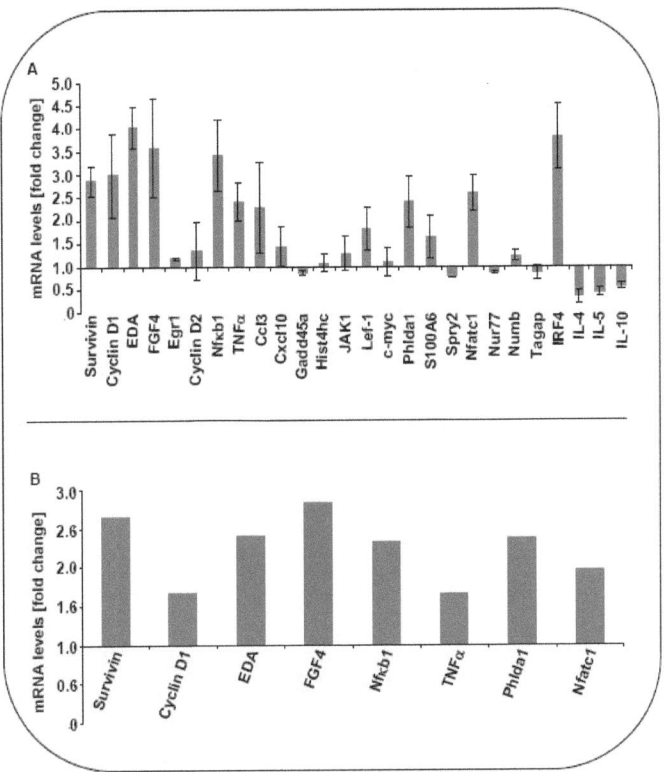

Figure 3.13 FasL reverse signaling target genes.
Splenic B cells were stimulated with 1 µg/ml anti-IgM antibody for four hours before isolation of total RNA and cDNA synthesis. Relative abundance of transcripts was quantified by qRT-PCR. Each sample was assessed in triplicate and normalized to *Hprt* (**A**) and *Gapdh* (**B**) as housekeeping genes. Transcript levels in *FasL ΔIntra* cells were related to expression levels in wildtype cells using the $2^{-\Delta\Delta Ct}$ method. (**A**) Columns represent mean values, bars represent the SEM of four independent experiments. (**B**) One experiment out of two is shown.

3.3.4.3 Significant regulation of Wnt signaling pathway-associated genes by the FasL ICD

In addition to proliferation- and activation-associated genes, transcript levels of several Lef-1 target genes, such as *Survivin*, *Cyclin D1*, *Ectodysplasin A* (*EDA*), *Fibroblast growth factor 4* (*FGF4*) and *Interleukin-4* (*IL-4*), were found to be affected in *FasL ΔIntra* B cells (**Fig. 3.13**). Based on this observation and on the previously demonstrated FasL ICD : Lef-1 interaction, expression of Wnt signaling pathway-related genes was investigated in more detail. A commercially available PCR array (mouse Wnt signaling pathway PCR array; *SABiosciences*) was used to compare the regulation of 86 Wnt pathway-related genes in activated B cells from *FasL ΔIntra* and wildtype mice. Surprisingly, a noticeable regulation was observed, with a significantly (≥ two-fold) reduced expression of 40 genes and a significantly elevated expression of one gene (*Cyclin D2*) (**Fig. 3.14**). Altogether, these data support a role for the FasL ICD in the Lef-1-dependent regulation of gene transcription.

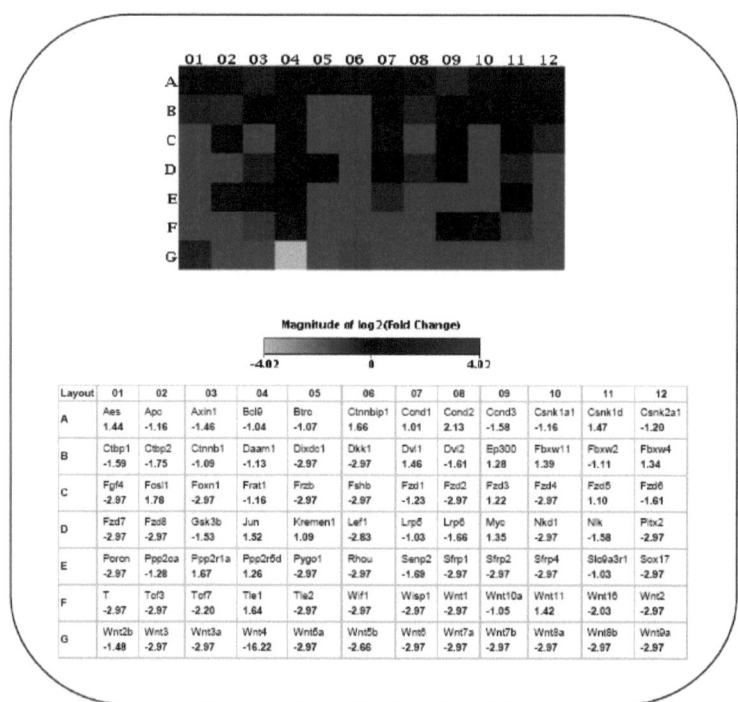

Figure 3.14 Extensive regulation of Wnt signaling pathway-associated genes.
The expression of 86 Wnt pathway-related genes in activated B cells (1 µg/ml anti-IgM antibody, four hours) was analyzed with the Mouse Wnt Signaling Pathway PCR Array from *SABiosciences*. Samples were normalized to six housekeeping genes. Relative mRNA levels were calculated using an appropriate web-based analysis tool (http://www.sabiosciences.com/pcr/arrayanalysis.php). The top panel shows a heat map of the fold change in *FasL ΔIntra* cells compared to wildtype cells. The bottom panel displays the genes analyzed and the respective fold change in *FasL ΔIntra* cells. Array analysis was performed once.

3.3.5 B cells isolated fom spleen express Lef-1

The expression of Lef-1 in B cells is controversially discussed. While Lef-1 expression in B cell progenitors (pre- and pro-B cells) is widely accepted, Lef-1 expression in mature B cells is debated (Reya et al., 2000). As the findings presented in this study strongly suggest an interplay of FasL ICD and Lef-1 in mature B cells, Lef-1 mRNA and protein expression were investigated in B cells isolated from the spleen of *FasL ΔIntra* or wildtype mice. Using conventional RT-PCR, the presence of *Lef-1* transcript in B cells from mutant and control mice could be demonstrated (**Fig. 3.15 A**). The proper identity of the PCR products obtained was supported by an identical signal derived from cDNA of HEK293T cells transfected with a Lef-1 expression plasmid. Likewise, intracellular staining with an anti-Lef-1 antibody for flow cytometric analysis revealed comparable amounts of Lef-1 protein in *FasL ΔIntra* and wildtype B cells (**Fig. 3.15 B**). These data indicate, that indeed Lef-1 mRNA and protein are expressed in mature B cells.

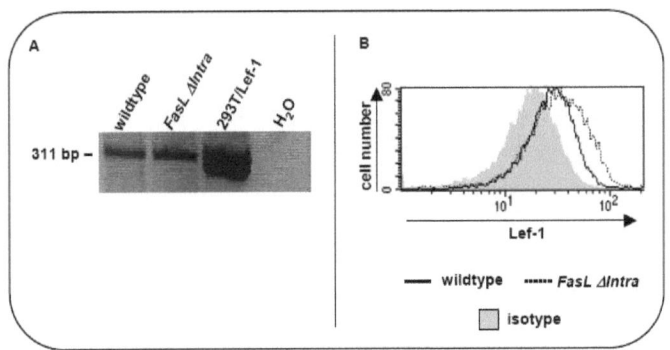

Figure 3.15 Lef-1 is expressed in splenic B cells.
(A) *Lef-1* mRNA expression in B cells of wildtype and *FasL ΔIntra* mice. B cells were isolated from spleen and stimulated with 1 µg/ml anti-IgM for four hours before RNA isolation and RT-PCR analysis using *Lef-1* specific primers. HEK293T cells transfected with a *Lef-1* expression construct served as positive control. **(B)** Lef-1 protein expression in B cells of wildtype and *FasL ΔIntra* mice. Freshly isolated splenic B cells were intracellularly stained for flow cytometric analysis using an anti-Lef-1 antibody and an Alexa Fluor546-coupled secondary antibody. One representative experiment out of six is shown. Wildtype cells are represented by the solid, *FasL ΔIntra* cells by the dotted line. The grey curve shows staining with an isotype control antibody.

To support this finding, Wnt signaling was investigated *in vivo*. Homozygous mice of the *FasL ΔIntra* mouse line were crossed to *TopGal* mice that express β-Galactosidase under the control of a Lef-1-dependent promoter (DasGupta *et al.*, 1999). Therefore, induction of Lef-1-dependent transcription correlates with β-Galactosidase activity, which can be quantified by flow cytometry as the extent of β-Galactosidase-mediated substrate conversion. For experiments, offspring bearing the *LacZ* transgene and either the wildtype or the truncated *FasL* locus was selected. B cells isolated from spleen were stimulated by BCR cross-linking for various time points before measurement of substrate conversion. Unfortunately, β-Galactosidase activity could never be detected (data not shown). This was not due to a general failure of the assay, as enzymatic activity could be detected reliably in B cells from Rosa mice that constitutively express a *LacZ* transgene.

3.4 *In vivo* studies to investigate the consequences of FasL reverse signaling

Hallmarks of an immune repsponse include the activation, recruitment and clonal expansion of innate and adaptive immune effector cells. Thus, changes in the proliferative capacity of lymphocytes, as observed in the absence of the FasL ICD *ex vivo*, should have an impact on the course of immune responses *in vivo*. Here, a potential participation of FasL reverse signaling in the regulation of immune responses to various challenges was investigated.

3.4.1 Analysis of thymocyte proliferation

Within the thymus, T cells develop from a double negative ($CD3^+CD4^-CD8^-$) to a double positive ($CD3^+CD4^+CD8^+$) and finally to a single positive (either $CD3^+CD4^+$ or $CD3^+CD8^+$) state. In order to reach the next maturational stage, thymocytes have to pass several selection steps to ensure that only T cells that function properly and do not recognize self-molecules are released into the periphery (Boursalian et al., 2003; Mueller, 2010). In a study from the group of Pamela Fink (Boursalian et al., 2003) FasL was reported to function as a co-stimulatory factor during positive selection. To investigate if T cell proliferation in the thymus is affeced by the absence or presence of the FasL ICD, mice were intraperitoneally injected with the thymidine analogue bromodesoxyuridine (BrdU). Co-staining of CD4, CD8 and BrdU revealed that BrdU was incorporated into the DNA of thymocytes of each developmental stage to the same extent in FasL ΔIntra and wildtype mice (**Fig. 3.16**). Together with the observation of normal mature T cell populations in the periphery of FasL ΔIntra mice (**Table 3.1**), this suggests, that the FasL ICD does not affect T cell development in the thymus.

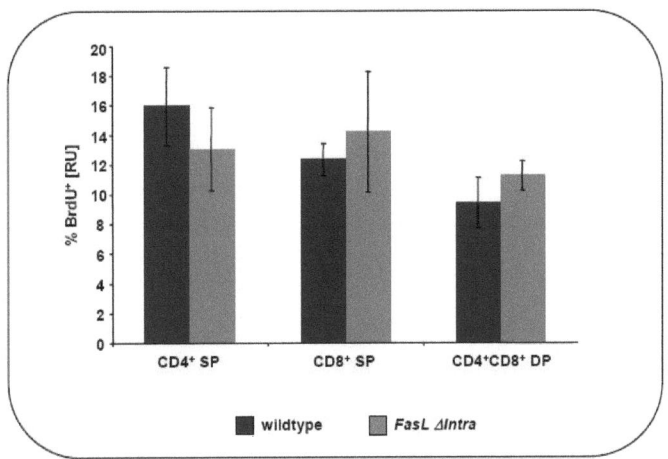

Figure 3.16 FasL ICD deletion does not affect thymocyte proliferation.
Three mice per group were injected i.p. twice with 1 mg BrdU/100 µl PBS (30 minutes delay). Control groups received PBS only. Six hours later, total thymocytes were stained for CD4 and CD8 cell surface expression and BrdU incoporation. The diagram shows the percentage of BrdU$^+$ cells in the depicted thymocyte populations normalized to untreated mice. Mean values and SEM of three experiments are shown. SP = single positive thymocytes; DP = double positive thymocytes

3.4.2 Expansion of Vβ8 T cell receptor (TCR) chain-expressing T cells

The proliferative capacity of T cells was analyzed in a simple *in vivo* model in which mice were intravenously injected with the bacterial superantigen *Staphylococcus* enterotoxin B (SEB). Mimicking a bacterial infection, SEB causes specifically the expansion of Vβ8 TCR chain-bearing T cells, whereas for example expansion of T cells with the Vβ6 TCR chain is not triggered (Desbarats *et al.*, 1998). In a three-color FACS analysis, the relative increase of helper (CD4$^+$) and cytotoxic (CD8$^+$) Vβ8- or Vβ6-expressing T cells was determined in spleen and lymph nodes of challenged *versus* untreated animals. As expected, a significant expansion of Vβ8$^+$ T cells, but not of Vβ6$^+$ T cells could be observed upon SEB administration (**Fig. 3.17**). However, the response of *FasL ΔIntra* mice matched that of wildtype mice. This indicates regulatory mechanisms operating *in vivo* that mask FasL reverse signaling, e.g. control exerted by regulatory T cells. Alternatively, the potent T cell stimulation by SEB might have overridden FasL reverse signaling which is only detectable under conditions of sub-optimal stimulation (Desbarats *et al.*, 1998; Suzuki *et al.*, 1998; Suzuki *et al.*, 2000b; Paulsen *et al.*, 2009).

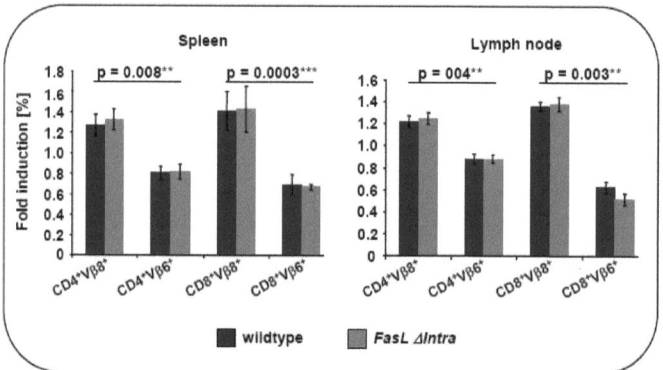

Figure 3.17 SEB-triggered expansion of Vβ8 TCR chain-expressing T cells.
Three mice per group were injected i.v. with 100 μg SEB. Total splenocytes and thymocytes were stained for cell surface expression of CD4, CD8 and Vβ8 or Vβ6 48 hours post-injection. Diagrams show the relative increase in Vβ8- or Vβ6-expressing T cells in challenged *vs.* unchallenged mice as fold induction. Columns represent the mean values and bars the SEM of three independent experiments.

3.4.3 Anti-viral response of CD8⁺ T cells following Lymphocytic choriomeningitis virus (LCMV) infection

LCMV belongs to the *arenaviridae* and is a widely used infection model to study various aspects of anti-viral immune reponses. LCMV-infection induces a protective response of virus-specific CD8⁺ cytotoxic T cells (CTLs), peaking after six to eight days, to clear the acute infection. In co-operation with Prof. Dr. Thomas Brunner (Bern, Switzerland), *FasL ΔIntra* and wildtype mice were intravenously infected with 10^3 PFU LCMV (strain WE) as previously described (Corazza *et al.*, 2000). The amount of virus-specific CD8⁺ T cells present in spleen was quantified by tetramer staining (epitope $H-2D^b$/gp33–41) eight days post-infection. At the peak of the anti-viral response, eight to nine percent of all CD8⁺ CTLs were virus-reactive (tetramer⁺) in both, *FasL ΔIntra* and wildtype animals (**Fig. 3.18**). Apparently, acute anti-viral CD8⁺ CTL responses to LCMV-infection are not regulated by the FasL ICD, or the strong immune stimulation by LCMV overrides reverse FasL signals.

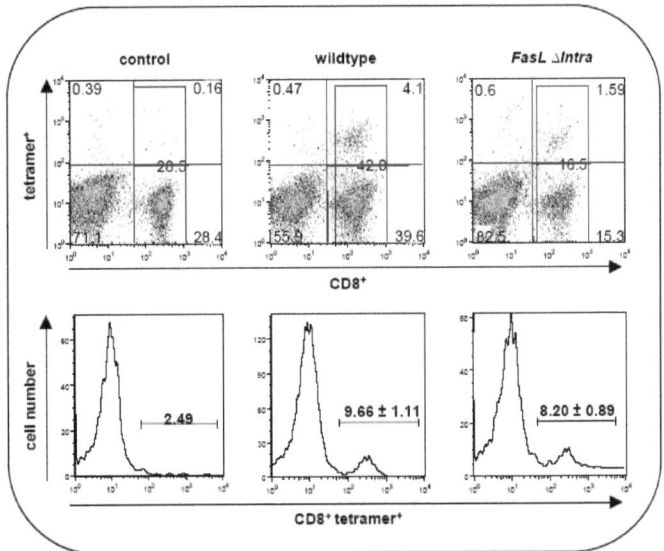

Figure 3.18 Antigen-specific expansion of CD8⁺ T cells after LCMV infection.
10^3 PFU LCMV-WE were injected i.v. into three mice per group. One control mouse was not infected. The amount of virus-specific CD8+ T cells present in spleen was quantified by tetramer staining (epitope $H-2D^b$/gp33-41) eight days post-infection.

3.4.4 Acute and long-term immunity in an infection model of listeriosis

Besides LCMV, infection with the intracellular bacterium *Listeria monocytogenes* is another well-described model to study antigen-specific T cell responses which clear the acute infection and constitute long-term immunological memory. Upon antigen encounter, antigen-specific naïve T cells expand clonally and acquire effector functions and the capacity to home to peripheral lymphoid tissues. Once the acute infection has been cleared, only a small subset of cells survive as antigen-specific memory cells with the capacity to rapidly respond to subsequent challenges.

With the help of the German Mouse Clinic (Dr. Thure Adler), acute and memory T cell responses towards infection with Ovalbumin (OVA)-expressing *L. monocytogenes* (*L.m.*-ova; strain 10403s) were analyzed in this study according to standard protocols (Busch *et al.*, 2001).

In a first experiment, the general capacity of *FasL ΔIntra* mice to cope with the bacterial burden was investigated. Therefore, mice were intravenously infected with a high dose (25,000) wildtype *L. monocytogenes*, provoking a non-specific inflammatory immune response within the first 72 hours. Livers and spleens from infected mice were prepared at day three, and the number of *Listeria* colony forming units (CFU) was calculated by plating dilutions of tissue homogenates on brain heart infusion (BHI) agar plates. As **Fig. 3.19** shows, high bacterial loads were present in spleen and liver of *FasL ΔIntra* and wildtype mice, indicating a comparable sensitivity to infection with *Listeria*.

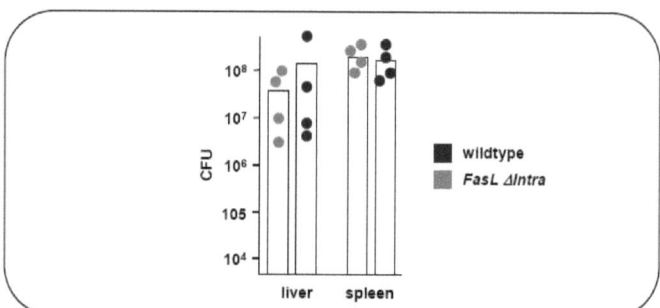

Figure 3.19 Sensitivity to infection with *Listeria*.
Four mice per group were injected i.v. with 25,000 wt *Listeria monocytogenes* (strain 10403s). Livers and spleens were prepared at day three post-infection. To assess the bacterial burden, *Listeria* CFU were calculated by plating dilutions of tissue homogenates on BHI agar plates.

To study acute antigen-specific immunity, mice were infected with a low dose (5,000) *L.m.*-ova, and the responses of epitope-reactive CD8$^+$ or CD4$^+$ T cells were characterized at day seven, *i.e.* at the peak of the primary T cell response. Tetramer stainings with the epitope H2-K^b/SIINFEKL revealed approximately four percent epitope-specific CD8$^+$ T cells in all animals (**Fig. 3.20 A, left**), the majority being effector T cells (~ 90% tetramer$^+$CD8$^+$ T cells were CD62L$^-$CD127$^-$) (**Fig. 3.20 A, right**). Clonal expansion correlated with the acquisition of effector functions in three to four percent of all CD4$^+$ or CD8$^+$ T cells, as indicated by the induction of cytokine production (**Fig. 3.20 B**). Levels of Interferon gamma (IFN-γ) and TNF-α-producing cells were quantified by intracellular stainings of cells *in vitro* stimulated for 5 hours with either the *Listeria* epitope (CD4$^+$ T cells: H2-K^b/LLO188-201; CD8$^+$ T cells: H2-K^b/SIINFEKL), DMSO (negative control) or PMA/ionomycine (positive control). Again, no difference in the number of cytokine-producing antigen-reactive CD4$^+$ or CD8$^+$ T cells was observed. Thus, the FasL ICD does not affect the sensitivity to acute infection with *Listeria* and apparently does not regulate the clonal expansion of bacteria-specific T cells or the acquisition of T cell effector functions.

The development of long-term immunity was investigated by sensitizing mice with a low dose of *L.m.*-ova (5,000) as before and subsequent challenge by a recall infection with a high dose *L.m.*-ova (250,000) ten weeks later, *i.e.* a long time period after clearance of the initial infection. Analogous to the primary infection, the sensitivity towards infection was assessed as the number of *Listeria* CFUs in spleen and liver homogenates two days after the challenge. While in naïve mice (not sensitized), the recall infection resulted in a high number of CFUs, the number of CFUs in spleen and liver of sensitized mice was almost below the detection limit in both, *FasL ΔIntra* and wildtype mice (**Fig. 3.21 A**). In agreement with the capacity of sensitized animals to rapidly cope with the bacterial burden, elevated numbers of *Listeria*-specific, cytokine-producing CD4$^+$ or CD8$^+$ T cells were found (**Fig. 3.21 B**, compare **Fig. 3.20**). These data suggest a normal development of immunologic memory in *FasL ΔIntra* mice following *Listeria*-infection.

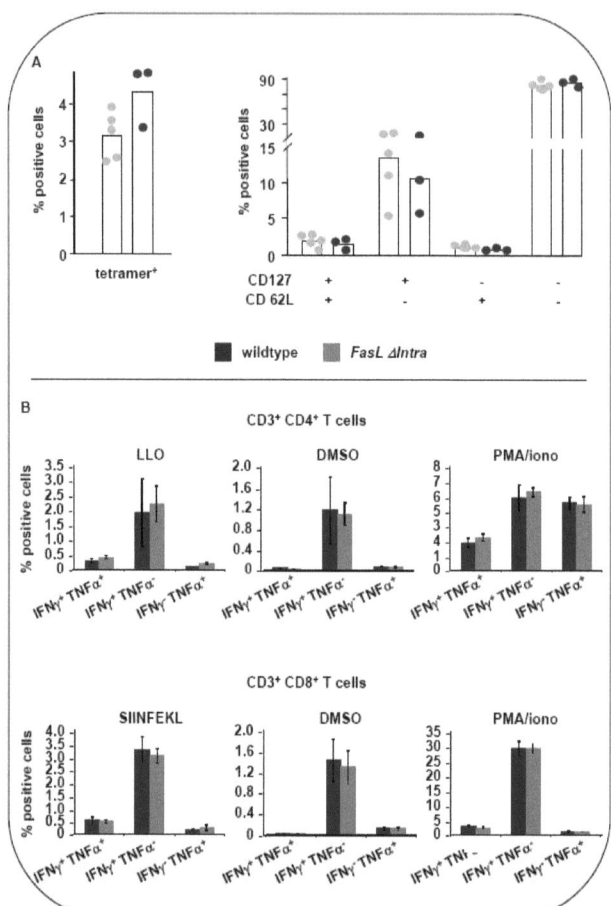

Figure 3.20 Analysis of the acute antigen-specific immunity following infection with *Listeria*. Five *FasL ΔIntra* mice and three wildtype littermates were injected i.v. with 5,000 *L.m.-ova*. Spleenocytes were prepared at day seven. **(A)** The epitope-specific CD8$^+$ T cell response was quantified by FACS analysis. left: Total amount of CD3$^+$CD8$^+$ tetramer$^+$ (H2-K^b/SIINFEKL) cells in the spleen of infected animals. right: Percentage of CD3$^+$CD8$^+$ tetramer$^+$ (H2-K^b/SIINFEKL) cells that express CD62L and/or CD127. **(B)** Intracellular cytokine staining for IFN-γ and TNF-α production. Cells were stimulated *in vitro* with either the *Listeria* epitope (CD4$^+$ T cells: H2-K^b/LLO188-201; CD8$^+$ T cells: H2-K^b/SIINFEKL), DMSO (negative control) or PMA/ionomycine (positive control). Subsequent flow cytometric analysis revealed the percentage of CD45$^+$ CD3$^+$, CD4$^+$ or CD8$^+$ lymphocytes producing IFN-γ and/or TNF-α.

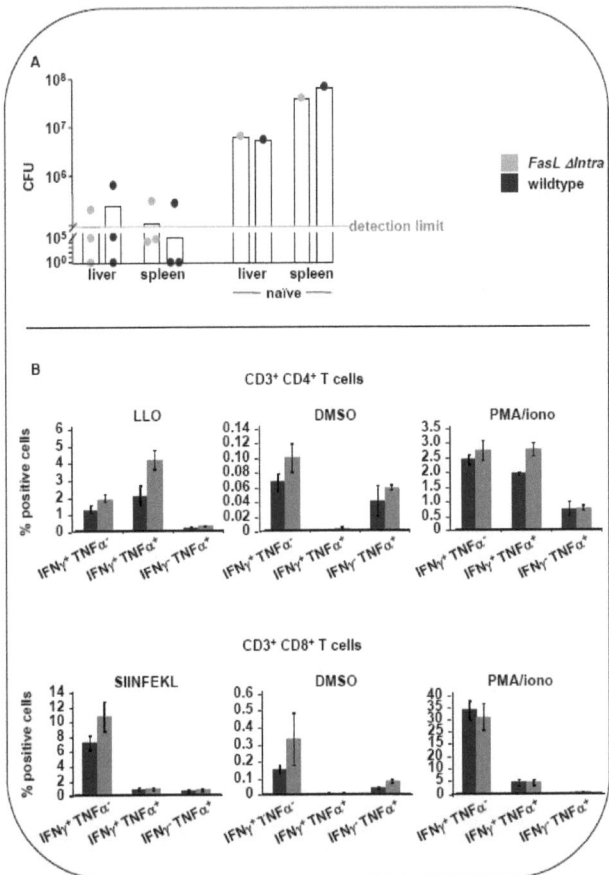

Figure 3.21 Long-term immunity following infection with *Listeria*.
Three mice per group were infected i.v. with 5,000 *L.m.-ova* and challenged with 250,000 *L.m.-ova* ten weeks later. One wildtype and one *FasL ΔIntra* mouse were not sensitized and received the second infection only (naïve). Livers and spleens were prepared two days after the recall infection. **(A)** Protection assay. *Listeria* CFU were calculated by plating dilutions of tissue homogenates on BHI agar plates. The diagram shows the bacterial burden in sensitized and naïve animals after the challenge. **(B)** Intracellular cytokine staining for IFN-γ and TNF-α production after *Listeria* challenge of sensitized mice. Splenocytes were stimulated *in vitro* with either the *Listeria* epitope (CD4$^+$ T cells: H2-K^b/LLO188-201; CD8$^+$ T cells: H2-K^b/SIINFEKL), DMSO (negative control) or PMA/ionomycine (positive control). Subsequent flow cytometric analysis revealed the percentage of CD45$^+$CD3$^+$, CD4$^+$ or CD8$^+$ lymphocytes producing IFN-γ and/or TNF-α.

3.4.5 Participation of the FasL ICD in Ovalbumin-induced allergic airway disease

In the initial phenotypic screen by the GMC, slightly elevated IgE levels were found in the serum of unchallenged *FasL ΔIntra* mice. This promted the use of an Ovalbumin (OVA)-induced model of allergic airway disease (AAD) in co-operation with the GMC. In this model, mice are sensitized to OVA peptide and later challenged by exposure to an OVA peptide-aerosol that leads to leukocyte recruitment to and their proliferation in the lung. Here, mice were sensitized with a low dose of alum-precipitated OVA peptide (OVA/alum; 10 µg on days zero and eight), hyposensitized with 200 µg OVA/alum on days 29 and 32 and finally challenged by exposure to OVA-aerosols (1% OVA peptide) on days 39, 42 and 45. The OVA peptide-induced overall number of branchoalveolar lavage (BAL)-infiltrating cells is very low, making analysis of sufficient cell numbers to obtain significant results difficult. Therefore, the hyposenzitation step was included into the protocol as it leads to enhanced cell recruitment into the BAL upon OVA-aerosol exposure (Aguilar-Pimentel *et al.*, 2010). After the final challenge, cell recruitment into the the BAL was determined by flow cytometry. Surprisingly, the total number of cells present in the BAL was significantly lower in *FasL ΔIntra* mice (**Fig. 3.22 A, left**). When analyzing the composition of cell populations within the BAL, it was revealed that significantly less B, T and NK cells had been recruited in the absence of the FasL ICD, while numbers of eosinophils, neutrophils and macrophages were similar (**Fig. 3.22 A, right**). At the same time, the overall percentage of activated $CD8^+$ T cells was significantly increased in the absence of the FasL ICD, as indicated by reduced levels of CD62L or Ly-6c-expressing $CD8^+$ T cells (**Fig. 3.22 B**). However, the activation status of antigen-specific (tetramer$^+$) $CD8^+$ T cells did not differ (data not shown). Together, these data suggest, that the enhanced activation of $CD8^+$ T cells in *FasL ΔIntra* mice attenuates the inflammatory reaction in this model of AAD by impairing leukocyte recruitment into the lung.

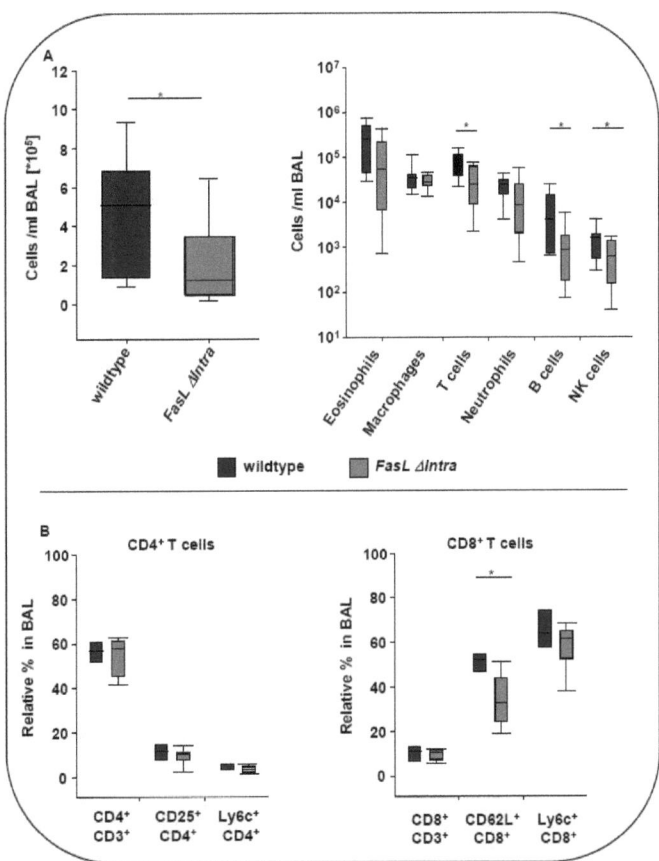

Figure 3.22 Participation of the FasL ICD in OVA peptide-induced allergic airway disease. Ten mice per group mice were sensitized with 10 μg OVA/alum each on days zero and eight, hyposensitized with 200 μg OVA/alum on days 29 and 32 and finally challenged by exposure to OVA-aerosols (1% OVA) on days 39, 42 and 45. Twentyfour hours after the final challenge, cell recruitment into the the branchoalveolar lavage (BAL) was determined by flow cytometry. **(A)** *left*: Significantly reduced cell recruitment into the BAL in *FasL ΔIntra* mice. The diagram displays the total number of cells per milliliter BAL. *right*: Significantly lower numbers of B, T and NK cells in the BAL of *FasL ΔIntra* mice. Differential cell count was performed by flow cytometry. **(B)** Enhanced activation of CD8$^+$ T cells in BAL of *FasL ΔIntra* mice. CD4$^+$ or CD8$^+$ T cells (living CD45$^+$CD3$^+$ cells) present in BAL were quantified (left most column in each diagram). Expression of activation markers in the respective CD4$^+$ or CD8$^+$ subpopulation was measured by FACS analysis by staining for Ly6c, CD25 or CD62L. CD62L expression negatively correlates with cell activation.

3.4.6 FasL reverse signaling modulates the germinal center reaction

Searching for an *in vivo* correlate for the enhanced B cell proliferation in the abence of the FasL ICD, a potential participation of FasL reverse signaling in thymus-dependent (TD) immune responses was investigated. Immunization with the hapten antigen 4-hydroxy-3-nitrophenyl acetate coupled to the carrier protein chicken gamma globuline (NP-CGG) results in the formation of germinal centers in the spleen. B cells recruited into germinal centers undergo somatic hypermutation and immunglobulin isotype class switch, resulting in plasma cells that produce high affinity antibodies predominantly of the IgG isotype. With help from Ursula Zimber-Strobel's group (Department of Gene Vectors, Helmholtz Zentrum München, Munich), germinal center reactions in *FasL ΔIntra* and wildtype mice were analyzed 14 days post-immunization. Flow cytometric analyses of splenocytes revealed significantly higher levels of germinal center B cells (GCs; $B220^+CD95^+PNA^+$) and of plasma cells (PCs; $Syndecan^+B220^{low}$ lymphocytes) in *FasL ΔIntra* mice following antigen exposure, providing an *in vivo* correlate for the enhanced B cell proliferation observed in these mice *ex vivo* (**Fig. 3.23 A**). At the same time, similar levels of marginal zone B cells (MzB) and of follicular B cells (FoB), *i.e.* of resting cells normally present in the spleen, were detected (**Fig. 3.23 B**).

Furthermore, immunization did not alter the overall amount of B and T cells in the spleen or the general activation status of B cells (**Table 3.3** and compare **Table 3.1**). This indicates that immunization did not lead to an unspecific, aberrant expansion of immune cells.

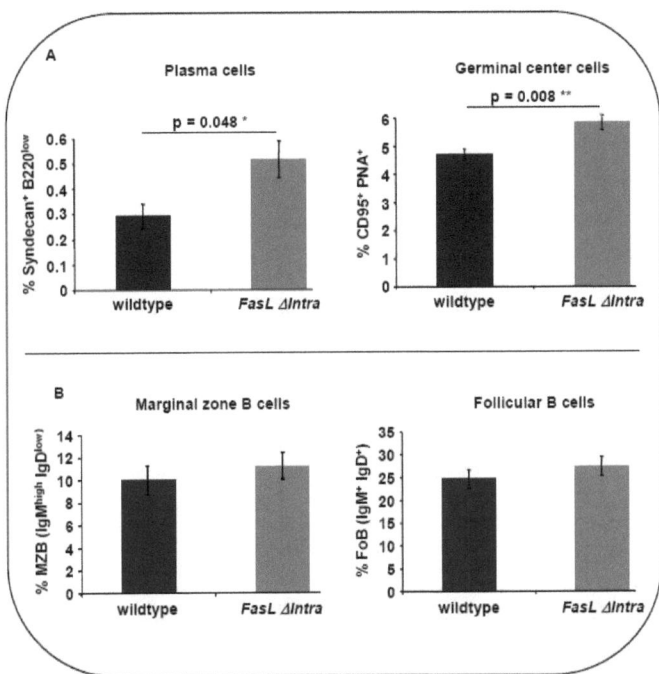

Figure 3.23 FasL reverse signaling modulates the germinal center reaction.
Mice were immunized by i.p. injection with 100 µg NP-CGG/alum and sacrificed after 14 days. Single cell suspensions of splenocytes were used for FACS analysis. **(A)** The amount of Syndecan$^+$ B220^{low} plasma cells (left) and CD95$^+$ PNA$^+$ germinal center B cells (right) was signifcantly higher in *FasL ΔIntra* mice. **(B)** Levels of marginal zone B cells (IgM^{high} IgD^{low}) and follicular B cells (IgM$^+$ IgD$^+$) are comparable in wildtype and *FasL ΔIntra* mice. Columns represent mean values and bars represent the SEM from two independent experiments with five mice per group each.

Table 3.3 Overall B cell and T cell levels in the spleen of immunized wildtype and *FasL ΔIntra* mice. Five mice per group were immunized by i.p. injection of 100 µg NP-CGG/alum. Fourteen days post-immunization, single cell suspensions of splenocytes were stained for the presence of B cells (B220$^+$) and T cells (CD3$^+$) and analyzed by flow cytometry. The activation status of B cells was determined by co-staining for CD80 and CD86.

	Marker profile	wildtype	*FasL ΔIntra*
T cells	CD3$^+$	26.47 ± 1.14	27.88 ± 1.63
B cells	B220$^+$	62.66 ± 1.63	61.86 ± 1.44
Activated B cells	B220$^+$CD80$^+$CD86$^+$	2.40 ± 0.13	2.29 ± 0.10

According to the enhanced germinal center reaction and the elevated number of PCs, serum titers of antigen-specific IgM and IgG antibodies should be increased in *FasL ΔIntra* mice. Following immunization with NP, antigen-specific antibodies are produced which can bind to the NP multimer NP17. Therefore, the levels of NP17-binding IgM and IgG1 antibodies in the serum of immunized mice were quantified by ELISA. Indeed, titers of both, IgM and IgG1 antibodies, were significantly higher in *FasL ΔIntra* than in wildtype mice after immunization (**Fig. 3.24**). Thus, the increased serum titers of NP-specific IgM and IgG1 antibodies correlated with the elevated number of PCs in *FasL ΔIntra* mice and support the notion that signaling *via* the FasL ICD apparently participates in the regulation of TD immune responses by dampening B cell activation and expansion in germinal centers.

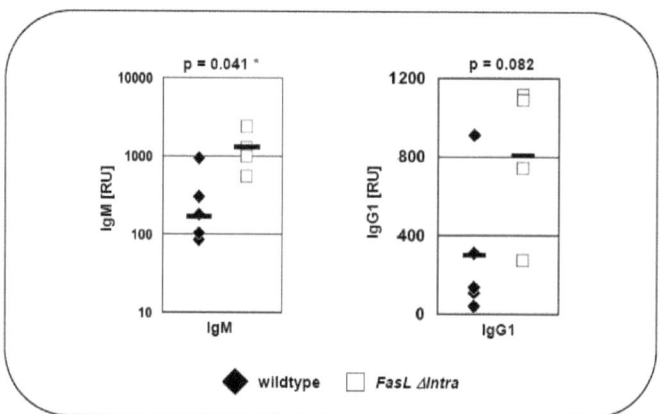

Figure 3.24 Significantly elevated titers of NP-specific IgM and IgG1 after immunization in *FasL ΔIntra* mice. Mice were immunized by i.p. injection with 100 µg NP-CGG/alum and sacrificed after 14 days. Serum was prepared from whole blood. The serum titers of NP17-specific IgM (left panel) and IgG1 (right panel) antibodies were quantified by ELISA. Each symbol represents one mouse, and the vertical bar indicates the mean value of five (wildtype) or four (*FasL ΔIntra*) animals per group. Immunglobulin titers in the serum of wildtype mice are represented by filled hash keys and titers in *FasL ΔIntra* mice by open rectangles.

3.4.7 Thioglycollate-induced peritonitis as a model for neutrophil migration

Recently, FasL expression on myeloid cells was reported to be important for recruitment of these cells to the site of infection. Notably, it was shown that both, induction of FasL cell surface expression and its cross-linking by Fas, were a prerequisit for migration (Letellier et al., 2010). However, myeloid cells can express FasL as well as Fas, and it has not been determined which signal, the FasL forward or the reverse signal, is the important one involved in this process.

We used thioglycollate-induced peritonitis to test an involvement of FasL reverse signaling in the recruitment of neutrophils. In mimickry of a systemic bacterial infection, thioglycollate injection into the peritoneal cavity induces peritonitis with the activation of innate immune responses and an increase of local neutrophils within the first hours. Following two challenges with thioglycollate, neutrophils (Ly6G$^+$) in the peritoneal lavage were quantified by flow cytometry three hours after the last challenge. While the induction of peritonitis resulted in a significant neutrophil recruitment (challenged vs. untreated) as expected, no differences between FasL $\Delta Intra$ and wildtype mice could be detected (**Fig. 3.25**). Thus, peritonitis-induced neutrophil recruitment is apparently not guided by a signal transduced via the FasL ICD.

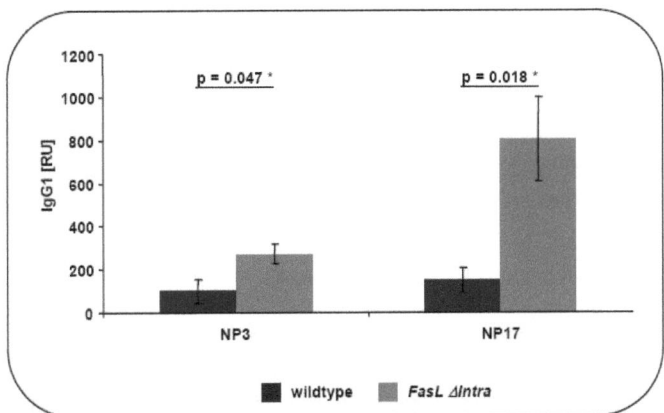

Figure 3.25 Neutrophil recruitment during thioglycollate-induced peritonitis.
Four mice per group were injected i.p. (+) with 1 ml [3%] thioglycollate broth in the evening of day one and were treated a second time in the morning of the following day. Control groups were left untreated (-). Neutrophil recruitment into the peritoneal lavage was determined by FACS analysis as the percentage of Ly6G (Gr-1)-expressing cells. Columns represent mean values and bars represent the SEM of data obtained from four mice each.

4 Discussion

Reverse signaling *via* the FasL ICD has been associated with the regulation of lymphocyte activation and proliferation (Lettau *et al.*, 2009), but conflicting data about the consequences exist, and neither the precise molecular mechanism, nor its physiological relevance have been addressed at the endogenous protein level *in vivo*. In the present study, a 'knockout/knockin' mouse model, in which wildtype *FasL* was replaced with a deletion mutant lacking the intracellular portion (*FasL ΔIntra*), was phenotypically characterized and was employed to investigate the physiological consequences of FasL reverse signaling at the molecular and cellular level.

4.1 *FasL ΔIntra* mice represent a suitable model for FasL reverse signaling deficiency and do not display an abnormal phenotype

To ensure that *FasL ΔIntra* mice represent a suitable model to study the consequences of FasL reverse signaling, FasL expression and functionality were analyzed. Proper execution of FasL/Fas-mediated apoptosis is an important issue as naturally occuring mutant mice, in which Fas-apoptosis is almost completely abrogated due to inactivating mutations in the FasL (*FasL$^{gld/gld}$*) and Fas (*Fas$^{lpr/lpr}$*) gene, respectively, develop a severe lymphoaccumulative syndrom with the occurrence of an aberrant CD3$^+$B220$^+$CD4$^-$CD8$^-$ T cell population, splenomegaly and auto-antibodies (Adachi *et al.*, 1993; Takahashi *et al.*, 1994a). A similar phenotype in *FasL ΔIntra* mice could mask the consequences of defective reverse signaling and, thus, render the mice inapt as an experimental model.

Comparable FasL expression on lymphocytes of homozygous *FasL ΔIntra* and wildtype mice was verified by RT-PCR and flow cytometric staining (**Fig. 3.3**). Co-culture assays demonstrate that the truncated FasL is still capable to induce apoptosis in Fas-bearing cells, although this is less efficient than with the full length protein (**Fig. 3.4**). This decrease in cytotoxicity is in agreement with a study showing that FasL raft partitioning is essential for maximum contact with the Fas receptor and its ability to induce Fas-mediated apoptosis. Raft localization of FasL is depending on the intracellular domain which is lacking in the *FasL ΔIntra* protein (Cahuzac *et al.*, 2006). However, the killing capacity of *FasL ΔIntra* effectors is significantly higher than that of cells derived from homozygous and heterozygous *FasL^{gld}* mice, of which only homozygous *FasL$^{gld/gld}$* animals develop the lymphoaccumulative diesase, while heterozygous $FasL^{wt/gld}$ mice are rescued by the presence of the single wildtype allele (Nagata et al., 1995).

Additionally, phenotypic screening of FasL ΔIntra mice does not reveal any phenotypic anomalies. Normal T and B cell populations develop, and even old mice do not present with any signs of the $FasL^{gld/gld}$ associated pathology (**Tables 3.1, 3.2** and data not shown). Notably, while $FasL^{gld/gld}$ cells express elevated levels of Fas at the surface (Stuart et al., 2005), normal Fas levels are found on cells isolated from FasL ΔIntra mice (**Fig. 3.7 B**). Although Fas expression is increased, $FasL^{gld/gld}$ T cell blasts fail to be eliminated by FasL-dependent AICD upon re-stimulation (Tanaka et al., 1995; O' Reilly et al., 2009). Thus, the comparable extent of cell death in lymphocytes from FasL ΔIntra and wildtype mice underlines the notion that FasL ΔIntra mice do not resemble FasL-defective $FasL^{gld/gld}$ or $FasL^{-/-}$ animals (**Fig. 3.6**).

The localization of FasL is thought to differ between hematopoietic and non-hematopoietic cells. In hematopoietic cells, FasL is stored intracellularly in multivesicular bodies and becomes externalized upon cell activation only, while non-hematopoietic cells constitutively express FasL at the cell surface (French et al., 1996a; Kavurma et al., 2003). It has been suggested that nascent FasL is sorted to secretory lysosomes (Bossi et al., 1999) and that this critically depends on the FasL PRD, since deletion of the cytoplasmic tail leads to the constitutive expression of FasL on the cell surface of hematopoietic cells (Blott et al., 2001). Probably, FasL sorting and targeted discharge at the immunological synapse is mediated and regulated by post-translational modifications and by the binding of cytosolic SH3 domain-containing proteins, such as Nck, to the PRD (Lettau et al., 2006; Qian et al., 2006; Zuccato et al., 2007). However, the non-pathologic phenotype of FasL ΔIntra mice does not support this hypothesis. Default FasL expression at the cell surface of lymphocytes, due to deletion of the ICD, should lead to the massive and systemic induction of apoptosis and severe disease. In contrast, expression levels and killing capacity of truncated and full length FasL are similar (**Fig. 3.3, 3.4**). It is not likely that this reflects a compensatory increased shedding of FasL from the surface of FasL ΔIntra cells, as membrane-bound FasL is the primary functional form of FasL in mice (O' Reilly et al., 2009). Furthermore, preliminary data indicate that the amount of FasL-containing microvesicles secreted by activated T cells into the culture medium is comparable in FasL ΔIntra and wildtype mice (data not shown). This agrees with results from Jurkat cells transfected with either wildtype or mutant (truncated ICD) FasL that secrete comparable amounts of vesicular FasL (Richard Siegel, personal communication). Providing possible explanations for this discrepancy, more recent studies claim that FasL is stored in compartments distinct from secretory lysosomes and suggest alternative pathways involving direct transport of FasL to the cell surface, not depending on the PRD (but rather on the total amount of FasL protein) (Xiao et al., 2004; He et al., 2007; Thornhill et al., 2007; Kassahn et al., 2009; Schmidt et al., 2009).

In summary, the *FasL ΔIntra* mouse line represents a suitable model to study the consequences of reverse signaling *via* the FasL ICD at the endogenous protein level *in vivo*, due to the proper expression of functional FasL protein (in terms of its apoptosis-inducing function).

4.2 Signaling *via* the FasL ICD impairs activation-induced lymphocyte proliferation by a mechanism involving PLCγ2, PKCα and ERK1/2

FasL reverse signaling has first been described by two studies that demonstrated either co-stimulation of murine $CD8^+$ T cell lines by FasL cross-linking or inhibition of activation-induced proliferation, IL-2 production and cell cycle progression of murine $CD4^+$ T cells. In both cases, co-stimulation or inhibition of TCR-initiated signals critically depended on the presence of signaling-competent FasL, as the capacity of $FasL^{gld/gld}$ T cells to expand in response to mitogenic stimulation differed from that of wildtype and $Fas^{lpr/lpr}$ cells (Desbarats et al., 1998; Suzuki et al., 1998). Based on these observations, the proliferative capacity of lymphocytes lacking the FasL ICD was investigated to decide which of these apparently contradictory findings, co-stimulation or inhibition, reflects the physiological role of FasL reverse signaling. It could be demonstrated that activation-induced proliferation of $CD4^+$ T cells, $CD8^+$ T cells and of B cells is enhanced in the absence of the FasL ICD (**Fig. 3.5**). Interestingly, this effect is most pronounced in B cells and could only be detected in $CD4^+$ T cells after depletion of $CD4^+CD25^+$ regulatory T cells. To our knowledge the findings presented here describe FasL reverse signaling in B cells for the first time.

These *ex vivo* data support a negative regulatory role for FasL in mitogen-induced expansion of lymphocytes. Consistently, the proliferative capacity of cells completely lacking FasL ($FasL^{-/-}$) is even higher than that of cells devoid of the FasL ICD only, a result that is also in agreement with current literature describing a proliferative advantage of $FasL^{-/-}$ cells over wildtype cells (Mabrouk et al., 2008).

In order to shed light on the molecules involved in transmission of the reverse FasL signal, the activation of several pathways that are known to play important roles in signaling from the antigen receptor was assessed. Providing a molecular correlate for the observed enhanced proliferation, ERK1/2 phosphorylation is significantly increased in B cells isolated from FasL ΔIntra mice upon BCR triggering (**Fig. 3.8**). Further characterization of the upstream pathway revealed an involvement of PLCγ2 and PKCα signals in the FasL-dependent ERK1/2 regulation: Absence of the FasL ICD augments the activating phosphorylation of PLCγ2 and PKCα, and pharmacologic inhibition of PKCα results in diminished ERK1/2 activation and attenuated proliferation, underlining an important role for PKCα in FasL reverse signaling (**Fig. 3.11**). Activation of PLCγ2 in B cells and of PLCγ1 in T cells by protein tyrosine kinases like Syk and Bruton's tyrosine kinase (Btk; B cells) or Fyn and Lymphocyte-specific protein tyrosine kinase (Lck; T cells) is a major downstream event of antigen receptor triggering (Kurosaki et al., 2000; Marshall et al., 2000). PLCγ then converts phosphatidylinositol 4,5-bisphosphate into the second messengers inositol 1,4,5-trisphosphate (IP3) and diacylglycerol (DAG). While IP3 promotes calcium release and NFAT activation, DAG is responsible for activation of PKC which in turn triggers the MAPK pathway.

Although PKCα has been reported to directly target c-Raf-1 and MEK1/2 (Kolch et al., 1993; Schonwasser et al., 1998), MAP-kinases upstream of ERK1/2 (Raf-1 and MEK1/2) seem not to be responsible for the differential regulation of ERK1/2 in our FasL mutant mice (**Fig. 3.9**). MEK1/2 inhibition impairs, but does not completely abolish B cell proliferation in response to BCR cross-linking, suggesting the involvement of an alternative, MAPK-cascade independent pathway in FasL ICD signaling. In agreement with this, PKC-dependent proliferation, specifically stimulated by PMA/ionomycine, is not affected by the presence of MEK inhibitors (**Fig. 3.10**).

The notion of PLCγ, PKCα and ERK1/2 as mediators of FasL reverse signaling is consistent with published data, demonstrating that concomitant engagement of FasL and the antigen receptor initiates reverse signaling via TCR-proximal molecules, such as ZAP-70, LAT, PLCγ, PKC and ERK1/2, and that cross-talk of the two signaling entities occurs at a very early step, probably at the level of receptor internalization (Sun et al., 2006; Paulsen et al., 2009). However, engagement of FasL enhances activating TCR signals according to Sun et al.

(2006), while it is inhibitory in primary human peripheral blood T cells analyzed by Paulsen and co-workers (2009).

How can the conflicting data regarding the consequences of FasL reverse signaling be explained? As detailed above, initial studies described either co-stimulatory or inhibitory functions in different murine $CD4^+$ or $CD8^+$ T cells. Follow-up studies by the group of Pamela Fink collected data supporting FasL-mediated co-stimulation and claimed that AICD masks the co-stimulatory signal in $CD4^+$ T cells in later phases of the immune response, thereby contradicting Desbarats et al. (1998) (Suzuki et al., 2000b; Boursalian et al., 2003; Sun et al., 2006; Sun et al., 2007b). A recent study on FasL reverse signaling challenges these reports and proposes an inhibitory FasL ICD/TCR cross-talk in human PBMCs (Paulsen et al., 2009). The discrepancies probably reflect the use of distinct and artificial experimental systems. In the ex vivo and in vivo experiments conducted with FasL ΔIntra mice during this project, FasL reverse signaling is initiated and regulated by its natural inducers, resulting in inhibitory signals. Other studies employed agonistic anti-FasL antibodies or Fas-Fc fusion proteins to cross-link FasL, possibly generating unphysiologically strong or weak signals. However, these reagents were used in both publications reporting co-stimulation and those describing inhibition mediated by FasL reverse signaling. Furthermore, Pamela Fink and co-workers used an artificial mouse FasL fusion construct which was overexpressed in murine $CD8^+$ T cell lines or human Jurkat cells (Sun et al., 2006; Sun et al., 2007b). This construct consists of the FasL ICD, dimerization domains that allow mimicry of ligand cross-linking, and a myristoylation motif for tethering the construct to the inner plasma membrane. A variant of this construct was engineered to harbor the transmembrane and extracellular domain of Ly49a, for cross-linking with an anti-Ly49a antibody, fused to the FasL ICD. Also in this model, the strength of FasL cross-linking and the extent of FasL triggering might be unphysiologically high.

It is important to note that effects observed in FasL ΔIntra mice indeed reflect FasL reverse signaling. First, the amount of cell death in stimulated and re-stimulated lymphocytes of FasL ΔIntra mice is comparable to that of wildtype mice (**Fig. 3.6**), showing that the differential proliferative response does not reflect diminished cell death. In addition, non-apoptotic/pro-survival signaling through the Fas receptor, which might account for the increased proliferation in cells devoid of the FasL ICD, could be excluded. Even in a Fas-defective setting ($FasL^{wt/wt}/Fas^{lpr/lpr}$ vs. $FasL^{\Delta/\Delta}/Fas^{lpr/lpr}$), the enhanced proliferative capacity of *FasL ΔIntra* lymphocytes is maintained (**Fig. 3.7**). Nevertheless, data from experiments employing $Fas^{lpr/lpr}$ or $FasL^{gld/gld}$ mice have to be interpreted with care, as neither mutation completely abrogates FasL/Fas signaling (Adachi *et al.*, 1993; Karray *et al.*, 2004; Mabrouk *et al.*, 2008). Furthermore, non-apoptotic Fas signals have been associated with co-stimulatory functions, so that the increased expansion of *FasL ΔIntra* cells, correlating with the absence of FasL reverse signaling, would have to be explained with increased pro-proliferative Fas engagement. There is no evidence supporting this hypothesis, as Fas cell surface expression levels are comparable in wildtype and *FasL ΔIntra* lymphocytes, and the latter exhibit a rather reduced killing capacity in co-culture assays (**Fig. 3.4**).

4.3 Reverse FasL signals regulate target gene transcription: differential expression of pro-proliferative and Lef-1-dependent genes in *FasL ΔIntra* mice

Previous studies in our group revealed a Notch-like proteolytic processing of FasL at the cell surface, resulting in the liberation of the FasL ICD after two consecutive cleavage steps. The liberated ICD was shown to translocate to the nucleus and regulate reporter gene expression (Kirkin *et al.*, 2007). Furthermore, using different *in vitro* protein binding assays, an interaction of the FasL ICD with the transcription factor Lef-1 could be demonstrated that leads to inhibition of Lef-1-dependent Luciferase activity (Lückerath *et al.*, manuscript submitted). Additionally, data presented in this study show a FasL ICD-mediated differential activation of ERK1/2, and ERK1/2 itself is involved in the transcriptional regulation of various target genes. Based on these findings, expression profiling of activated lymphocytes was performed to identify *bona fide* FasL reverse signaling target genes. The enhanced proliferation in the absence of the FasL ICD correlates with and can be explained by the upregulation of several genes associated with lymphocyte proliferation and activation in homozygous *FasL ΔIntra* mice relative to wildtype mice, for example of *NFAT*, *NF-κb* and *Irf4* (**Fig. 3.13**).

Interestingly, an extensive regulation of genes related to Wnt/β-Catenin- and/or Lef-1-signaling, such as *Cyclin D1*, *EDA*, *FGF4* and *IL-4*, was found as well (**Fig. 3.13, 3.14**). Consistent with the repression of *IL-4* transcription by direct binding of Lef-1 to the *IL-4* promoter (Hebenstreit *et al.*, 2008), *IL-4* mRNA levels are significantly decreased in *FasL ΔIntra* B cells, where Lef-1 activity is enhanced due to the absence of the inhibiting FasL ICD. FasL expression in mature B cells has been confirmed in several studies (Hahne *et al.*, 1996;

Lundy et al., 2002; Lundy, 2009; Lundy et al., 2009), but different publications suggest that Lef-1 expression becomes downregulated during B cell development (Reya et al., 2000; Jin et al., 2002). Contrary to these reports, Lef-1 mRNA and protein can be detected in splenic B cells isolated from *FasL ΔIntra* and wildtype mice (**Fig. 3.15**). Co-expression of both, FasL and Lef-1, allows FasL ICD-mediated inhibition of Lef-1-dependent gene expression in mature lymphocytes, initiated by Notch-like processing of FasL.

However, direct interaction of the FasL ICD with Lef-1 has only been shown *in vitro*, and Wnt/Lef-1 signaling could not be detected *in vivo* in splenic B cells from $FasL^{wt/wt}/TopGal$ and $FasL^{\Delta/\Delta}/TopGal$ mice (data not shown). *TopGal* mice, similar to *BatGal* mice, express the *LacZ* gene under control of a Lef-1-dependent promoter. These mouse lines do not allow sensitive detection of Wnt signaling, and the absence of a positive signal does not exclude Wnt/Lef-1 activity (Barolo, 2006). Thus, from our experiments it cannot be concluded if the failure to detect Luciferase signals in the reporter mice is due to the insensitivity of the experimental model or to the lack of Lef-1 activity.

4.4 Molecular model for FasL reverse signaling

Based on the data presented here, a model is proposed in which reverse signals transmitted by the FasL ICD proceed through at least two different pathways to regulate early events of antigen receptor signaling, resulting in transcriptional regulation of target genes and, finally, in attenuation of lymphocyte proliferation (**Fig. 4.1**).

(i) We previously reported a ternary complex consisting of full length FasL, the SH3 domain-containing adaptor PSTPIP and the phosphatase PTP-PEST, whose formation is promoted by FasL cross-linking (Baum et al., 2005). This complex is an attractive candidate for the initiation of FasL reverse signaling. Upon concomitant antigen receptor and FasL engagement, PSTPIP binds to the FasL PRD and recruits PTP-PEST, resulting in enhanced phosphatase acitivty. PTP-PEST, associated with PSTPIP, has been described as a negative regulator of lymphocyte activation (Davidson et al., 2001; Veillette et al., 2002; Yang et al., 2006; Arimura et al., 2008). As such, overexpression of the human analog of PSTPIP, CD2BP1, in the B cell lymphoma cell line A20 selectively diminished the phosphorylation of PLCγ, ERK1/2 and p38 upon antigen receptor cross-linking in a PTP-PEST-dependent manner (Yang et al., 2006). Furthermore, several PTP-PEST substrates have been identified, including Lck, Shc, Cytoplasmic tyrosine kinase (Csk), Grb2, Breast cancer antiestrogen resistance 1 (BCAR1; Cas), Focal adhesion kinase (FAK), and Proline-rich tyrosine kinase 2 (Pyk 2), that are involved in antigen receptor proximal signaling events and might directly or indirectly regulate PLCγ activation (Davidson et al., 2001; Faisal et al., 2002; Yamamoto et al., 2003; Arimura et al., 2008; Veillette et al., 2009). Thus, FasL reverse signals could be transmitted through a FasL/PSTPIP/PTP-PEST complex that forms upon FasL triggering and in which PTP-PEST activity is increased. This allows cross-talk with antigen receptor-initiated pathways at an early step, resulting in diminished PLCγ, PKC and ERK1/2 activation and finally attenuating the transcription of pro-proliferative genes and the expansion of lymphocytes. Interestingly, PTP-PEST has recently been described to sensitize cells to death receptor-mediated apoptosis. While PTP-PEST enhances caspase activity during apoptosis initiation, it subsequently becomes cleaved by Caspase-3 itself. This cleavage regulates the catalytic activity, interactions and adaptor functions of PTP-PEST to ultimately facilitate the characteristic rounding up of dying cells (Halle et al., 2007). The anti-proliferative and pro-apoptotic effects of FasL/PSTPIP/PTP-PEST might possibly work hand in hand in the termination of immune responses by simultaneously preventing ongoing clonal expansion and promoting AICD of lymphocytes.

However, other scenarios how reverse signaling via full length FasL can regulate lymphocyte proliferation can be envisioned as well. PSTPIP/PTP-PEST have been implicated in the regulation of T cell activation through cytoskeletal rearrangements. By binding of PSTPIP/PTP-PEST to the cytoplasmic tail of CD2, molecules involved in the re-organization of the Actin cytoskeleton are recruited to enable formation of the immunological synapse and, hence, optimal T cell activation (Li et al., 1998; Bai et al., 2001; Badour et al., 2003). A similar mechanism could be possible for the interaction of FasL, PSTPIP and PTP-PEST. Furthermore, it cannot be excluded that FasL reverse signaling regulates PLCγ2 activity by a completely different mechanism, for example through the adaptor proteins Grb2 or Gab1, both of which have been identified as FasL PRD interaction partners that play important roles in BCR-induced PLCγ activation (Marshall et al., 2000; Pao et al., 2007; Voss et al., 2009). Likewise, a proteomic screen identified the IL-2-inducible T cell kinase (Itk) as a potential interaction partner of the FasL PRD (Voss et al., 2009). Analysis of $Itk^{-/-}$ mice indicates an important role of Itk in TCR-induced PLCγ phosphorylation and in the promotion of T cell proliferation (Berg et al., 2005). To elucidate how FasL reverse signaling modulates PLCγ activity, knockdown experiments could be performed in lymphocytes isolated from wildtype and mutant mice in which the expression of PTP-PEST or an alternative protein is silenced, for example, by virally-transduced shRNAs.

(ii) An alternative pathway for FasL reverse signaling is initiated *via* Notch-like processing of cell surface FasL by the proteases ADAM10 and SPPL2a, which leads to liberation of the FasL ICD into the cytoplasm and its translocation into the nucleus (Kirkin *et al.*, 2007). In the nucleus, the FasL ICD can bind to Lef-1 to negatively regulate Lef-1-dependent transcription, predominantely of pro-proliferative target genes. Although a direct FasL ICD/Lef-1 interaction has been shown *in vitro*, the exact molecular mechanism how the FasL ICD inhibits Lef-1 transcriptional activity remains unclear. Lef-1 binding seems to require the first 39 aa in the FasL ICD and, consistent with the lack of a SH3 domain in Lef-1, appears independent of the FasL PRD. Regarding the binding of FasL to Lef-1, a direct competition of the FasL ICD with the Lef-1 activating β-Catenin is unlikely, as the predicted binding site for FasL does not overlap with the N-terminal β-Catenin binding site in Lef-1 (Lückerath *et al.*, manuscript submitted and unpublished data).

In light of the important role of Lef-1 in B cell development, it could be assumed that the FasL ICD/Lef-1 interaction does not exclusively occur in mature lymphocytes, but also in B cell progenitors to prime B cells. In agreement with this, direct binding of a Lef-1-containing complex to the *Recombination-activating gene-2* (*Rag-2*) promoter in immature B cells influences *V(D)J* gene recombination, resulting in differential immunglobulin expression in mature B cells (Jin *et al.*, 2002). To test this possibility, *Rag-2* expression levels could be quantified and patterns of *V(D)J* gene recombination could be analyzed in B cell progenitors isolated from *FasL ΔIntra* and wildtype mice.

There might also be an additional or alternative route for the regulation of Lef-1 by FasL reverse signaling, not involving direct interaction of the two proteins. The FasL ICD-dependent PLCγ and PKC regulation could lead to modulation of Glycogen synthase kinase-3 beta (GSK-3β) activity by PKC, resulting in the accumulation of β-Catenin and the activation of Lef-1 (Christian *et al.*, 2002).

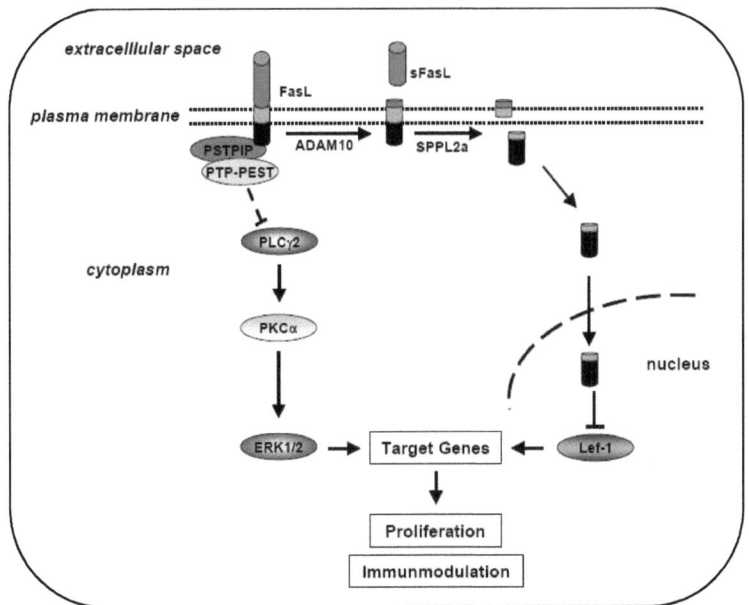

Figure 4.1 Molecular model for FasL reverse signaling.
FasL reverse signaling negatively regulates lymphocyte proliferation by at least two different pathways: FasL cross-linking promotes the formation of a ternary complex consisting of FasL, PSTPIP and PTP-PEST that enhances the phosphatase activity of PTP-PEST. As a consequence, PLCγ, PKC and ERK1/2 activation is inhibited, resulting in diminished transcription of pro-proliferative genes. Alternatively (and not necessarily exclusively), proteolysis of FasL liberates the ICD that might translocate to the nucelus to bind Lef-1 and, thus, to inhibit Lef-1-dependent transcription of pro-proliferative and Wnt signaling-related genes.

It would be interesting to identify the factors that determine which of these two pathways is triggered. A possibility would be the differential post-translational modification of the FasL ICD, such as phosphorylation of the tyrosine residue at position seven *vs.* serine phosphorylation within the CK-I site or *vs.* C82 palmitoylation which has already been identified as a pre-requisite for FasL raft localization and proteolysis by ADAM10 (Guardiola *et al.*, manuscript in preparation). Likewise, it is intriguing to speculate about a potential interplay between the proposed PLCγ/PKC/ERK1/2- and the FasL ICD/Lef-1-dependent pathways: Lef-1 activity could be affected by PKC-mediated inhibition of GSK-3β; Wnt/Lef-1 signaling can regulate and/or feedback on PLCγ and PKC *via* the Wnt-Ca^{2+}-pathway, in which Wnt agonists trigger the Ca^{2+}-dependent activation of PLCγ and PKC (Staal *et al.*, 2008).

4.5 Analysis of FasL reverse signaling *in vivo*

A potential participation of the FasL ICD in the regulation of immune responses to various challenges was analyzed *in vivo*, to elucidate the function of FasL reverse signaling in the context of the whole organism. Immune responses are characterized by the activation, recruitment and clonal expansion of immune effector cells. Thus, modulation of the proliferative capacity of lymphocytes, as observed *ex vivo*, should impact their course *in vivo*.

4.5.1 Signaling *via* the FasL ICD is apparently not important for thymocyte development

FasL reverse signaling has been reported to aid in the positive selection of thymocytes in OT-1 TCR transgenic mice (Boursalian *et al.*, 2003). To investigate whether T cell proliferation and maturation in the thymus of *FasL ΔIntra* mice is disturbed in the absence of the FasL ICD, BrdU incorporation into newly synthesized DNA was quantified (**Fig. 3.16**). In contrast to the results from Boursalian and co-workers, the findings presented here do not support a role for FasL signaling in T cell development, since neither thymic nor peripheral T cell subsets of *FasL ΔIntra* mice deviated from wildtype cells. Most likely, this discrepancy reflects the use of different experimental systems. In the study by Boursalian *et al.* (2003), $FasL^{wt/wt}$ and $FasL^{gld/gld}$ mice expressing either a diverse, polyclonal TCR repertoir (C57Bl/6N) or a transgenic TCR specific for an OVA-derived epitope (OT-1/C57Bl/6N) were compared. Strikingly, not only OT-1/$FasL^{gld/gld}$ but also OT-1/$FasL^{wt/wt}$ mice displayed a lower thymic cellularity, an abnormal distribution of thymocyte subsets and an increased expression of FasL when compared to standard C57Bl/6N mice. Furthermore, while substantial differences in the relative abundance of thymic T cell populations of OT-1/$FasL^{gld/gld}$ and OT-1/$FasL^{wt/wt}$ were found, BrdU incorporation was not affected by the $FasL^{gld/gld}$ mutation. Thus, FasL reverse signaling might aid positive selection under very specific circumstances only and, in agreement with the findings in *FasL ΔIntra* mice, does not seem to play a major role during thymocyte selection.

However, the data presented in this study should be interpreted with caution: double positive (DP) T cells represent the most abundant thymic T cell population (Oehme et al., 2005), but appear as the least abundant subset here.

4.5.2 FasL reverse signaling modulates the germinal center reaction

Encounter of forgein antigens leads to the clonal expansion and differentiation of non-circulating marginal zone B cells (MzBs) and circulating follicular B cells (FoBs) in the periphery. FoBs form germinal centers in which they undergo class switch recombination of immunglobulins, affinity maturation and further proliferation to give rise to short lived, antibody secreting plasma cells (PCs). Differentiation into PCs is guided by the transcription factors B lymphocyte-induced maturation protein 1 (Blimp-1), X box-binding protein-1 (Xbp-1) and IRF-4 (McHeyzer-Williams et al., 2005; Tarlinton et al., 2008). In agreement with the observed hyperactivation of lymphocytes in the absence of the ICD ex vivo, in vivo immunization with a TD-antigen revealed elevated numbers of PCs and a significantly increased generation of germinal center B cells (GCs) in response to the immunization in homozygous FasL ΔIntra mice (**Fig. 3.23**), leading to increased titers of NP-binding, antigen-specific IgM and IgG1 antibodies in the serum (**Fig. 3.24**). Possibly, the upregulation of Irf4 mRNA levels in activated B cells of FasL ΔIntra mice (**Fig. 3.13**) might contribute to the higher percentages of PCs in the spleen. At the same time, numbers of MzBs and FoBs, of overall T and B cell subsets and of activated B cells are comparable before and after immunization (**Fig. 3.23** and **Tables 3.1, 3.3**), indicating a specific response. These in vivo data match the previous ex vivo findings, and suggest that FasL reverse signaling exerts immunmodulatory functions.

Interestingly, β-Catenin knockout mice present with a similar phenotype. While T and B cell populations are normal, in vitro immunization experiments reveal a defect in the generation of PCs that correlates with decreased IRF-4 levels and a slight enhancement of class switch recombination to IgG3 and IgG1 (Yu et al., 2008b). Thus, FasL ICD signaling through Lef-1, possibly involving the modulation of GSK-3β activity and hence, of β-Catenin levels, might contribute to the germinal center reaction in FasL ΔIntra mice. An involvement of Lef-1 is supported by increased Irf-4 expression in the absence of the FasL ICD (**Fig. 3.13**), as Lef-1 may regulate Irf-4 by direct binding to a consensus binding site within the Irf-4 regulatory region (Yu et al., 2008b).

4.5.3 Signals transmitted through the FasL ICD participate in Ovalbumin-induced allergic airway disease

Supporting an immunomodulatory function of FasL reverse signaling, a participation of FasL in the course of Ovalbumin-induced allergic airway inflammation (AAD) was detected. Upon re-challenge of sensitized animals with OVA, significantly less cells are recruited into the BAL of *FasL ΔIntra* mice. Differential cell counts within the BAL reveal that absolute cell numbers of all cell types analyzed (eosinophils, macrophages, neutrophils, T cells, B cells, NK cells) are slightly decreased, with T, B and NK cells being significantly underrepresented in *FasL ΔIntra* mice, although their relative abundance is maintained (**Fig. 3.22** and data not shown). At the same time, activation of $CD8^+$ T cells ($CD62L^-$) is enhanced in the absence of the FasL ICD as measured by CD62L staining. This is largely consistent with a study in which enhanced activation of antigen-specific $CD8^+$ T cells negatively correlates with disease severity, *i.e.* cell recruitment into the BAL, eosinophilia and serum IgE levels (Aguilar-Pimentel *et al.*, 2010), and with the hyperactivation of *FasL ΔIntra* lymphocytes observed *ex vivo* and *in vivo*. Surprisingly, the percentage and activation status of antigen-specific $CD8^+$ T cells is similar in wildtype and *FasL ΔIntra* mice (data not shown) and, hence, the beneficial effect during the course of AAD would need to be explained by enhanced activation of unspecific $CD8^+$ T cells. Several possible explanations for this discrepancy have been suggested: (i) The OVA-challenge might have been not strong enough to reveal relevant differences in antigen-specific effector cells; (ii) The propensity to undergo AICD might be altered; (iii) Homing defects might lead to the sequestration of antigen-specific lymphocytes in other organs, e.g. in peripheral lymph nodes or the liver (Aguilar-Pimentel *et al.*, 2010); Marcus Ollert, personal communication). However, the analysis of AICD in *FasL ΔIntra* and wildtype T and B cells *ex vivo* does not provide evidence for a differential sensitivity to cell death (**Fig. 3.6**). It is clear, that the analysis performed with this OVA peptide-induced model of AAD does not allow elucidation of the underlying mechanism, and further experiments are needed to solve the contradictory findings.

From the presented findings, it can already be deduced that FasL reverse signaling participates in the course of OVA-induced AAD, particularly in the recruitment of immune effector cells into the lung. Indeed, a role for FasL signals in the migration and recruitment of neutrophils and macrophages to the site of inflammation has recently been described (Letellier *et al.*, 2010). Furthermore, $FasL^{gld/gld}$ mice were found to be less sensitive to chemically-induced colitis compared to wildype or $Fas^{lpr/lpr}$ mice, and FasL/Fas signaling competence correlated with severity of inflammation (Park et al., 2010). In contrast, in a mouse model of spontaneous coloncarcinoma ($Apc^{Min/+}$), the $FasL^{gld}$ mutation was associated with enhanced tumor formation and reduced neutrophil infiltration into inflamed or cancerous tissue (Fingleton et al., 2007). In a collaboration with Dr. Benno Weigmann (University of Erlangen) we are currently investigating the susceptibiltiy of FasL ΔIntra mice to chemically-induced colitis and colon cancer. Preliminary data suggest a lower tumor score in the absence of the FasL ICD. It will be interesting to see which cells infiltrate the inflamed and the cancerous tissue, and wether FasL reverse signaling contributes to cell recruitment and/or effector functions in this model.

It is attractive to speculate that PTP-PEST may execute these reverse FasL signals: PTP-PEST has been implicated in the modulation of lymphocyte activation and of cytoskeletal re-organization and is an important positive regulator of cell migration in fibroblasts (Veillette et al., 2009). Thus, a ternary FasL/PSTPIP/PTP-PEST complex might have an impact on immune cell migration, infiltration and/or other processes involving cytoskeletal re-organization, like granule exocytosis (i.e. effector functions).

4.5.4 Regulatory mechanism operating *in vivo* mask FasL reverse signaling

Several *in vivo* experiments have been conducted in which no significant difference between FasL ΔIntra and wildtype mice became apparent. These findings might reflect regulatory mechanisms operating *in vivo* to fine-tune immune responses, such as control exerted by regulatory T cells. Along these lines, proliferative differences in $CD4^+$ T cells could only be detected *ex vivo* after depletion of $CD4^+CD25^+$ regulatory T cells (**Fig. 3.5**). Support for the notion of a regulatory T cell-mediated suppression of $CD4^+$ T cell proliferation comes from studies in which no obvious difference in the expansion of Vβ8 TCR chain-expressing $CD4^+$ T cells following administration of the superantigen SEB was observed (**Fig. 3.17**). The comparable extent of SEB-induced $Vβ8^+$ $CD8^+$ T cell expansion in FasL ΔIntra and wildtype mice has been expected, since neither the $FasL^{gld/gld}$ nor the $Fas^{lpr/lpr}$ mutation affect the *in vivo* proliferation of this T cell population (Desbarats et al., 1998).

Furthermore, several *in vitro* studies indicate that FasL reverse signal-transduction can only be observed under conditions of sub-optimal lymphocyte stimulation (Desbarats et al., 1998; Suzuki et al., 1998; Boursalian et al., 2003; Sun et al., 2007b; Paulsen et al., 2009). If TCR signaling is induced with high (optimal) stimuli doses, FasL cross-linking has no detectable impact. A likely explanation for this phenomenon is the postulated modulating nature of the reverse FasL signal during antigen receptor-triggered lymphocyte activation. Strong receptor signals may override FasL signals and/or make them redundant, while weak signals might rely on the modulation exerted by FasL. Thus, the requirement for sub-optimal antigen receptor stimulation to reveal consequences of FasL reverse signaling offers an alternative or additional explanation for the comparable results obtained in experiments which analyzed either the expansion of antigen-specific T cells following a challenge with the superantigen SEB, with LCMV or with *L. monocytogenes*, or the recruitment of neutrophils in a thioglycollate-induced model of peritonitis (**Fig. 3.17-21; Fig. 3.25**). All of these challenging stimuli are potent activators of immune responses with the potential to mask inhibitory FasL reverse signaling (Desbarats *et al.*, 1998; Busch *et al.*, 2001; Huster *et al.*, 2006; Hegazy *et al.*, 2010; Letellier *et al.*, 2010).

4.6 Conclusion

In the study presented, *FasL ΔIntra* mice, which express a truncated FasL lacking the ICD, are characterized and analyzed as a model for FasL reverse signaling deficiency to elucidate the molecular mechanism and physiological importance of retrograde signal-transduction *via* the FasL ICD. While no obvious phenotypic anomalies can be detected in unchallenged *FasL ΔIntra* mice, *ex vivo* lymphocyte activation is augmented in the absence of reverse FasL signals as revealed by an enhanced activation-induced proliferation of B and T cells. This is mediated by amplification of antigen receptor proximal events, such as increased phosphorylation of PLCγ, PKC and ERK1/2, and by upregulation of pro-proliferative genes. The hyperactivation of cells isolated from *FasL ΔIntra* mice correlates with increased expansion and/or activation of lymphocytes in *in vivo* models of germinal center reactions and allergic airway inflammation and suggests an immunomodulatory function of FasL reverse signaling. However, regulatory mechanisms acting *in vivo* probably mask retrograde signals in other immune challenge experiments.

In summary, the here reported *ex vivo* and *in vivo* findings based on endogenous FasL protein levels demonstrate that FasL ICD-mediated reverse signaling is a negative modulator of certain immune responses. It is tempting to speculate that FasL reverse signaling might be a fine-tuning mechanism to prevent autoimmune diseases, a possibility which will be tested in the future with suitable auto-immune mouse models.

5 Literature

Adachi, M., Watanabe-Fukunaga, R., Nagata, S. (1993). Aberrant transcription caused by the insertion of an early transposable element in an intron of the Fas antigen gene of lpr mice. *Proc Natl Acad Sci U S A* 90(5): 1756-1760.

Aggarwal, B. B. (2003). Signalling pathways of the TNF superfamily: a double-edged sword. *Nat Rev Immunol* 3(9): 745-756.

Aguilar-Pimentel, J. A., Alessandrini, F., Huster, K. M., Jakob, T., Schulz, H., Behrendt, H., Ring, J., de Angelis, M. H., Busch, D. H., Mempel, M., Ollert, M. (2010). Specific CD8 T cells in IgE-mediated allergy correlate with allergen dose and allergic phenotype. *Am J Respir Crit Care Med* 181(1): 7-16.

Algeciras-Schimnich, A., Shen, L., Barnhart, B. C., Murmann, A. E., Burkhardt, J. K., Peter, M. E. (2002). Molecular ordering of the initial signaling events of CD95. *Mol Cell Biol* 22(1): 207-220.

Allison, J., Thomas, H. E., Catterall, T., Kay, T. W., Strasser, A. (2005). Transgenic expression of dominant-negative Fas-associated death domain protein in beta cells protects against Fas ligand-induced apoptosis and reduces spontaneous diabetes in nonobese diabetic mice. *J Immunol* 175(1): 293-301.

Ametller, E., Garcia-Recio, S., Costamagna, D., Mayordomo, C., Fernandez-Nogueira, P., Carbo, N., Pastor-Arroyo, E. M., Gascon, P., Almendro, V. (2010). Tumor promoting effects of CD95 signaling in chemoresistant cells. *Mol Cancer* 9(1): 161.

Andera, L. (2009). Signaling activated by the death receptors of the TNFR family. *Biomed Pap Med Fac Univ Palacky Olomouc Czech Repub* 153(3): 173-180.

Arimura, Y., Vang, T., Tautz, L., Williams, S., Mustelin, T. (2008). TCR-induced downregulation of protein tyrosine phosphatase PEST augments secondary T cell responses. *Mol Immunol* 45(11): 3074-3084.

Badour, K., Zhang, J., Shi, F., McGavin, M. K., Rampersad, V., Hardy, L. A., Field, D., Siminovitch, K. A. (2003). The Wiskott-Aldrich syndrome protein acts downstream of CD2 and the CD2AP and PSTPIP1 adaptors to promote formation of the immunological synapse. *Immunity* 18(1): 141-154.

Bai, Y., Ding, Y., Spencer, S., Lasky, L. A., Bromberg, J. S. (2001). Regulation of the association between PSTPIP and CD2 in murine T cells. *Exp Mol Pathol* 71(2): 115-124.

Barnhart, B. C., Legembre, P., Pietras, E., Bubici, C., Franzoso, G., Peter, M. E. (2004). CD95 ligand induces motility and invasiveness of apoptosis-resistant tumor cells. *EMBO J* 23(15): 3175-3185.

Barolo, S. (2006). Transgenic Wnt/TCF pathway reporters: all you need is Lef? *Oncogene* 25(57): 7505-7511.

Baum, W., Kirkin, V., Fernandez, S. B., Pick, R., Lettau, M., Janssen, O., Zornig, M. (2005). Binding of the intracellular Fas ligand (FasL) domain to the adaptor protein PSTPIP results in a cytoplasmic localization of FasL. *J Biol Chem* 280(48): 40012-40024.

Bellgrau, D., Gold, D., Selawry, H., Moore, J., Franzusoff, A., Duke, R. C. (1995). A role for CD95 ligand in preventing graft rejection. *Nature* 377(6550): 630-632.

Bentele, M., Lavrik, I., Ulrich, M., Stosser, S., Heermann, D. W., Kalthoff, H., Krammer, P. H., Eils, R. (2004). Mathematical modeling reveals threshold mechanism in CD95-induced apoptosis. *J Cell Biol* 166(6): 839-851.

Berg, L. J., Finkelstein, L. D., Lucas, J. A., Schwartzberg, P. L. (2005). Tec family kinases in T lymphocyte development and function. *Annu Rev Immunol* 23: 549-600.

Blott, E. J., Bossi, G., Clark, R., Zvelebil, M., Griffiths, G. M. (2001). Fas ligand is targeted to secretory lysosomes via a proline-rich domain in its cytoplasmic tail. *J Cell Sci* 114(Pt 13): 2405-2416.

Bodmer, J. L., Schneider, P., Tschopp, J. (2002). The molecular architecture of the TNF superfamily. *Trends Biochem Sci* 27(1): 19-26.

Bonardelle, D., Benihoud, K., Kiger, N., Bobe, P. (2005). B lymphocytes mediate Fas-dependent cytotoxicity in MRL/lpr mice. *J Leukoc Biol* 78(5): 1052-1059.

Bossi, G. and Griffiths, G. M. (1999). Degranulation plays an essential part in regulating cell surface expression of Fas ligand in T cells and natural killer cells. *Nat Med* 5(1): 90-96.

Boursalian, T. E. and Fink, P. J. (2003). Mutation in fas ligand impairs maturation of thymocytes bearing moderate affinity T cell receptors. *J Exp Med* 198(2): 349-360.

Brenner, D., Krammer, P. H., Arnold, R. (2008). Concepts of activated T cell death. *Crit Rev Oncol Hematol* 66(1): 52-64.

Brunner, T., Arnold, D., Wasem, C., Herren, S., Frutschi, C. (2001). Regulation of cell death and survival in intestinal intraepithelial lymphocytes. *Cell Death Differ* 8(7): 706-714.

Bulfone-Paus, S., Bulanova, E., Budagian, V., Paus, R. (2006). The interleukin-15/interleukin-15 receptor system as a model for juxtacrine and reverse signaling. *Bioessays* 28(4): 362-377.

Busch, D. H., Vijh, S., Pamer, E. G. (2001). Animal model for infection with Listeria monocytogenes. *Curr Protoc Immunol* Chapter 19: Unit 19 19.

Cahuzac, N., Baum, W., Kirkin, V., Conchonaud, F., Wawrezinieck, L., Marguet, D., Janssen, O., Zornig, M., Hueber, A. O. (2006). Fas ligand is localized to membrane rafts, where it displays increased cell death-inducing activity. *Blood* 107(6): 2384-2391.

Chakrabandhu, K., Herincs, Z., Huault, S., Dost, B., Peng, L., Conchonaud, F., Marguet, D., He, H. T., Hueber, A. O. (2007). Palmitoylation is required for efficient Fas cell death signaling. *EMBO J* 26(1): 209-220.

Chakrabandhu, K., Huault, S., Garmy, N., Fantini, J., Stebe, E., Mailfert, S., Marguet, D., Hueber, A. O. (2008). The extracellular glycosphingolipid-binding motif of Fas defines its internalization route, mode and outcome of signals upon activation by ligand. *Cell Death Differ* 15(12): 1824-1837.

Chan, F. K., Chun, H. J., Zheng, L., Siegel, R. M., Bui, K. L., Lenardo, M. J. (2000). A domain in TNF receptors that mediates ligand-independent receptor assembly and signaling. *Science* 288(5475): 2351-2354.

Chen, L., Park, S. M., Tumanov, A. V., Hau, A., Sawada, K., Feig, C., Turner, J. R., Fu, Y. X., Romero, I. L., Lengyel, E., Peter, M. E. (2010). CD95 promotes tumour growth. *Nature* 465(7297): 492-496.

Christian, S. L., Sims, P. V., Gold, M. R. (2002). The B cell antigen receptor regulates the transcriptional activator beta-catenin via protein kinase C-mediated inhibition of glycogen synthase kinase-3. *J Immunol* 169(2): 758-769.

Chu, J. L., Drappa, J., Parnassa, A., Elkon, K. B. (1993). The defect in Fas mRNA expression in MRL/lpr mice is associated with insertion of the retrotransposon, ETn. *J Exp Med* 178(2): 723-730.

Cohen, P. L. and Eisenberg, R. A. (1991). Lpr and gld: single gene models of systemic autoimmunity and lymphoproliferative disease. *Annu Rev Immunol* 9: 243-269.

Corazza, N., Muller, S., Brunner, T., Kagi, D., Mueller, C. (2000). Differential contribution of Fas- and perforin-mediated mechanisms to the cell-mediated cytotoxic activity of naive and in vivo-primed intestinal intraepithelial lymphocytes. *J Immunol* 164(1): 398-403.

Corsini, N. S., Sancho-Martinez, I., Laudenklos, S., Glagow, D., Kumar, S., Letellier, E., Koch, P., Teodorczyk, M., Kleber, S., Klussmann, S., Wiestler, B., Brustle, O., Mueller, W.,

Gieffers, C., Hill, O., Thiemann, M., Seedorf, M., Gretz, N., Sprengel, R., Celikel, T., Martin-Villalba, A. (2009). The death receptor CD95 activates adult neural stem cells for working memory formation and brain repair. *Cell Stem Cell* 5(2): 178-190.

D'Souza, B., Miyamoto, A., Weinmaster, G. (2008). The many facets of Notch ligands. *Oncogene* 27(38): 5148-5167.

DasGupta, R. and Fuchs, E. (1999). Multiple roles for activated LEF/TCF transcription complexes during hair follicle development and differentiation. *Development* 126(20): 4557-4568.

Davidson, D. and Veillette, A. (2001). PTP-PEST, a scaffold protein tyrosine phosphatase, negatively regulates lymphocyte activation by targeting a unique set of substrates. *EMBO J* 20(13): 3414-3426.

Davidson, W. F., Giese, T., Fredrickson, T. N. (1998). Spontaneous development of plasmacytoid tumors in mice with defective Fas-Fas ligand interactions. *J Exp Med* 187(11): 1825-1838.

Desbarats, J., Duke, R. C., Newell, M. K. (1998). Newly discovered role for Fas ligand in the cell-cycle arrest of CD4+ T cells. *Nat Med* 4(12): 1377-1382.

Dohrman, A., Russell, J. Q., Cuenin, S., Fortner, K., Tschopp, J., Budd, R. C. (2005). Cellular FLIP long form augments caspase activity and death of T cells through heterodimerization with and activation of caspase-8. *J Immunol* 175(1): 311-318.

Ehrenschwender, M. and Wajant, H. (2009). The Role of FasL and Fas in Health and Disease. *Adv Exp Med Biol* 647: 64-93.

Eissner, G., Kolch, W., Scheurich, P. (2004). Ligands working as receptors: reverse signaling by members of the TNF superfamily enhance the plasticity of the immune system. *Cytokine Growth Factor Rev* 15(5): 353-366.

Faisal, A., el-Shemerly, M., Hess, D., Nagamine, Y. (2002). Serine/threonine phosphorylation of ShcA. Regulation of protein-tyrosine phosphatase-pest binding and involvement in insulin signaling. *J Biol Chem* 277(33): 30144-30152.

Fas, S. C., Fritzsching, B., Suri-Payer, E., Krammer, P. H. (2006). Death receptor signaling and its function in the immune system. *Curr Dir Autoimmun* 9: 1-17.

Fingleton, B., Carter, K. J., Matrisian, L. M. (2007). Loss of functional Fas ligand enhances intestinal tumorigenesis in the Min mouse model. *Cancer Res* 67(10): 4800-4806.

French, L. E., Hahne, M., Viard, I., Radlgruber, G., Zanone, R., Becker, K., Muller, C., Tschopp, J. (1996a). Fas and Fas ligand in embryos and adult mice: ligand expression in several immune-privileged tissues and coexpression in adult tissues characterized by apoptotic cell turnover. *J Cell Biol* 133(2): 335-343.

French, L. E. and Tschopp, J. (1996b). Constitutive Fas ligand expression in several non-lymphoid mouse tissues: implications for immune-protection and cell turnover. *Behring Inst Mitt*(97): 156-160.

Gajate, C. and Mollinedo, F. (2005). Cytoskeleton-mediated death receptor and ligand concentration in lipid rafts forms apoptosis-promoting clusters in cancer chemotherapy. *J Biol Chem* 280(12): 11641-11647.

Green, D. R., Droin, N., Pinkoski, M. (2003). Activation-induced cell death in T cells. *Immunol Rev* 193: 70-81.

Green, D. R. and Ferguson, T. A. (2001). The role of Fas ligand in immune privilege. *Nat Rev Mol Cell Biol* 2(12): 917-924.

Griffith, T. S., Brunner, T., Fletcher, S. M., Green, D. R., Ferguson, T. A. (1995). Fas ligand-induced apoptosis as a mechanism of immune privilege. *Science* 270(5239): 1189-1192.

Guicciardi, M. E. and Gores, G. J. (2009). Life and death by death receptors. *FASEB J* 23(6): 1625-1637.

Guzman-Rojas, L., Sims-Mourtada, J. C., Rangel, R., Martinez-Valdez, H. (2002). Life and death within germinal centres: a double-edged sword. *Immunology* 107(2): 167-175.

Hahne, M., Peitsch, M. C., Irmler, M., Schroter, M., Lowin, B., Rousseau, M., Bron, C., Renno, T., French, L., Tschopp, J. (1995). Characterization of the non-functional Fas ligand of gld mice. *Int Immunol* 7(9): 1381-1386.

Hahne, M., Renno, T., Schroeter, M., Irmler, M., French, L., Bornard, T., MacDonald, H. R., Tschopp, J. (1996). Activated B cells express functional Fas ligand. *Eur J Immunol* 26(3): 721-724.

Halle, M., Liu, Y. C., Hardy, S., Theberge, J. F., Blanchetot, C., Bourdeau, A., Meng, T. C., Tremblay, M. L. (2007). Caspase-3 regulates catalytic activity and scaffolding functions of the protein tyrosine phosphatase PEST, a novel modulator of the apoptotic response. *Mol Cell Biol* 27(3): 1172-1190.

He, J. S., Gong, D. E., Ostergaard, H. L. (2010). Stored Fas ligand, a mediator of rapid CTL-mediated killing, has a lower threshold for response than degranulation or newly synthesized Fas ligand. *J Immunol* 184(2): 555-563.

He, J. S. and Ostergaard, H. L. (2007). CTLs contain and use intracellular stores of FasL distinct from cytolytic granules. *J Immunol* 179(4): 2339-2348.

Hebenstreit, D., Giaisi, M., Treiber, M. K., Zhang, X. B., Mi, H. F., Horejs-Hoeck, J., Andersen, K. G., Krammer, P. H., Duschl, A., Li-Weber, M. (2008). LEF-1 negatively controls interleukin-4 expression through a proximal promoter regulatory element. *J Biol Chem* 283(33): 22490-22497.

Hegazy, A. N., Peine, M., Helmstetter, C., Panse, I., Frohlich, A., Bergthaler, A., Flatz, L., Pinschewer, D. D., Radbruch, A., Lohning, M. (2010). Interferons direct Th2 cell reprogramming to generate a stable GATA-3(+)T-bet(+) cell subset with combined Th2 and Th1 cell functions. *Immunity* 32(1): 116-128.

Hildeman, D. A., Zhu, Y., Mitchell, T. C., Bouillet, P., Strasser, A., Kappler, J., Marrack, P. (2002). Activated T cell death in vivo mediated by proapoptotic bcl-2 family member bim. *Immunity* 16(6): 759-767.

Hotchkiss, R. S., Strasser, A., McDunn, J. E., Swanson, P. E. (2009). Cell death. *N Engl J Med* 361(16): 1570-1583.

Hu, W. H., Johnson, H., Shu, H. B. (2000). Activation of NF-kappaB by FADD, Casper, and caspase-8. *J Biol Chem* 275(15): 10838-10844.

Hueber, A. O., Bernard, A. M., Herincs, Z., Couzinet, A., He, H. T. (2002). An essential role for membrane rafts in the initiation of Fas/CD95-triggered cell death in mouse thymocytes. *EMBO Rep* 3(2): 190-196.

Huster, K. M., Stemberger, C., Busch, D. H. (2006). Protective immunity towards intracellular pathogens. *Curr Opin Immunol* 18(4): 458-464.

Igney, F. H. and Krammer, P. H. (2005). Tumor counterattack: fact or fiction? *Cancer Immunol Immunother* 54(11): 1127-1136.

Itoh, M., Chen, X. H., Takeuchi, Y., Miki, T. (1995). Morphological demonstration of the immune privilege in the testis using adjuvants: tissue responses of male reproductive organs in mice injected with Bordetella pertussigens. *Arch Histol Cytol* 58(5): 575-579.

Janssen, O., Qian, J., Linkermann, A., Kabelitz, D. (2003). CD95 ligand--death factor and costimulatory molecule? *Cell Death Differ* 10(11): 1215-1225.

Jin, Z. and El-Deiry, W. S. (2005). Overview of cell death signaling pathways. *Cancer Biol Ther* 4(2): 139-163.

Jin, Z. X., Kishi, H., Wei, X. C., Matsuda, T., Saito, S., Muraguchi, A. (2002). Lymphoid enhancer-binding factor-1 binds and activates the recombination-activating gene-2 promoter together with c-Myb and Pax-5 in immature B cells. *J Immunol* 169(7): 3783-3792.

Karray, S., Kress, C., Cuvellier, S., Hue-Beauvais, C., Damotte, D., Babinet, C., Levi-Strauss, M. (2004). Complete loss of Fas ligand gene causes massive lymphoproliferation and early death, indicating a residual activity of gld allele. *J Immunol* 172(4): 2118-2125.

Kassahn, D., Nachbur, U., Conus, S., Micheau, O., Schneider, P., Simon, H. U., Brunner, T. (2009). Distinct requirements for activation-induced cell surface expression of preformed Fas/CD95 ligand and cytolytic granule markers in T cells. *Cell Death Differ* 16(1): 115-124.

Kavurma, M. M. and Khachigian, L. M. (2003). Signaling and transcriptional control of Fas ligand gene expression. *Cell Death Differ* 10(1): 36-44.

Kirkin, V., Cahuzac, N., Guardiola-Serrano, F., Huault, S., Luckerath, K., Friedmann, E., Novac, N., Wels, W. S., Martoglio, B., Hueber, A. O., Zornig, M. (2007). The Fas ligand intracellular domain is released by ADAM10 and SPPL2a cleavage in T-cells. *Cell Death Differ* 14(9): 1678-1687.

Knox, P. G., Milner, A. E., Green, N. K., Eliopoulos, A. G., Young, L. S. (2003). Inhibition of metalloproteinase cleavage enhances the cytotoxicity of Fas ligand. *J Immunol* 170(2): 677-685.

Kolch, W., Heidecker, G., Kochs, G., Hummel, R., Vahidi, H., Mischak, H., Finkenzeller, G., Marme, D., Rapp, U. R. (1993). Protein kinase C alpha activates RAF-1 by direct phosphorylation. *Nature* 364(6434): 249-252.

Koncz, G., Kerekes, K., Chakrabandhu, K., Hueber, A. O. (2008). Regulating Vav1 phosphorylation by the SHP-1 tyrosine phosphatase is a fine-tuning mechanism for the negative regulation of DISC formation and Fas-mediated cell death signaling. *Cell Death Differ* 15(3): 494-503.

Krammer, P. H., Arnold, R., Lavrik, I. N. (2007). Life and death in peripheral T cells. *Nat Rev Immunol* 7(7): 532-542.

Krishna, M. and Narang, H. (2008). The complexity of mitogen-activated protein kinases (MAPKs) made simple. *Cell Mol Life Sci* 65(22): 3525-3544.

Kurosaki, T., Maeda, A., Ishiai, M., Hashimoto, A., Inabe, K., Takata, M. (2000). Regulation of the phospholipase C-gamma2 pathway in B cells. *Immunol Rev* 176: 19-29.

Lavrik, I. N., Golks, A., Riess, D., Bentele, M., Eils, R., Krammer, P. H. (2007). Analysis of CD95 threshold signaling: triggering of CD95 (FAS/APO-1) at low concentrations primarily results in survival signaling. *J Biol Chem* 282(18): 13664-13671.

Lee, K. H., Feig, C., Tchikov, V., Schickel, R., Hallas, C., Schutze, S., Peter, M. E., Chan, A. C. (2006). The role of receptor internalization in CD95 signaling. *EMBO J* 25(5): 1009-1023.

Letellier, E., Kumar, S., Sancho-Martinez, I., Krauth, S., Funke-Kaiser, A., Laudenklos, S., Konecki, K., Klussmann, S., Corsini, N. S., Kleber, S., Drost, N., Neumann, A., Levi-Strauss, M., Brors, B., Gretz, N., Edler, L., Fischer, C., Hill, O., Thiemann, M., Biglari, B., Karray, S., Martin-Villalba, A. (2010). CD95-ligand on peripheral myeloid cells activates Syk kinase to trigger their recruitment to the inflammatory site. *Immunity* 32(2): 240-252.

Lettau, L. A. (2004). Universal precautions studios presents: ID creature features. *Clin Infect Dis* 38(7): 1043.

Lettau, M., Paulsen, M., Kabelitz, D., Janssen, O. (2009). FasL expression and reverse signalling. *Results Probl Cell Differ* 49: 49-61.

Lettau, M., Qian, J., Linkermann, A., Latreille, M., Larose, L., Kabelitz, D., Janssen, O. (2006). The adaptor protein Nck interacts with Fas ligand: Guiding the death factor to the cytotoxic immunological synapse. *Proc Natl Acad Sci U S A* 103(15): 5911-5916.

Li, J., Nishizawa, K., An, W., Hussey, R. E., Lialios, F. E., Salgia, R., Sunder-Plassmann, R., Reinherz, E. L. (1998). A cdc15-like adaptor protein (CD2BP1) interacts with the CD2 cytoplasmic domain and regulates CD2-triggered adhesion. *EMBO J* 17(24): 7320-7336.

Lundy, S. K. (2009). Killer B lymphocytes: the evidence and the potential. *Inflamm Res*.

Lundy, S. K. and Boros, D. L. (2002). Fas ligand-expressing B-1a lymphocytes mediate CD4(+)-T-cell apoptosis during schistosomal infection: induction by interleukin 4 (IL-4) and IL-10. *Infect Immun* 70(2): 812-819.

Lundy, S. K. and Fox, D. A. (2009). Reduced Fas ligand-expressing splenic CD5+ B lymphocytes in severe collagen-induced arthritis. *Arthritis Res Ther* 11(4): R128.

Mabrouk, I., Buart, S., Hasmim, M., Michiels, C., Connault, E., Opolon, P., Chiocchia, G., Levi-Strauss, M., Chouaib, S., Karray, S. (2008). Prevention of autoimmunity and control of recall response to exogenous antigen by Fas death receptor ligand expression on T cells. *Immunity* 29(6): 922-933.

Maetzel, D., Denzel, S., Mack, B., Canis, M., Went, P., Benk, M., Kieu, C., Papior, P., Baeuerle, P. A., Munz, M., Gires, O. (2009). Nuclear signalling by tumour-associated antigen EpCAM. *Nat Cell Biol* 11(2): 162-171.

Marshall, A. J., Niiro, H., Yun, T. J., Clark, E. A. (2000). Regulation of B-cell activation and differentiation by the phosphatidylinositol 3-kinase and phospholipase Cgamma pathway. *Immunol Rev* 176: 30-46.

Martinez-Lorenzo, M. J., Anel, A., Gamen, S., Monle n, I., Lasierra, P., Larrad, L., Pineiro, A., Alava, M. A., Naval, J. (1999). Activated human T cells release bioactive Fas ligand and APO2 ligand in microvesicles. *J Immunol* 163(3): 1274-1281.

Matozaki, T., Murata, Y., Okazawa, H., Ohnishi, H. (2009). Functions and molecular mechanisms of the CD47-SIRPalpha signalling pathway. *Trends Cell Biol* 19(2): 72-80.

Matsumoto, N., Imamura, R., Suda, T. (2007). Caspase-8- and JNK-dependent AP-1 activation is required for Fas ligand-induced IL-8 production. *FEBS J* 274(9): 2376-2384.

Matthies, K. M., Newman, J. L., Hodzic, A., Wingett, D. G. (2006). Differential regulation of soluble and membrane CD40L proteins in T cells. *Cell Immunol* 241(1): 47-58.

McHeyzer-Williams, L. J. and McHeyzer-Williams, M. G. (2005). Antigen-specific memory B cell development. *Annu Rev Immunol* 23: 487-513.

McHeyzer-Williams, M. G. (2003). B cells as effectors. *Curr Opin Immunol* 15(3): 354-361.

Mizuno, T., Zhong, X., Rothstein, T. L. (2003). Fas-induced apoptosis in B cells. *Apoptosis* 8(5): 451-460.

Mueller, D. L. (2010). Mechanisms maintaining peripheral tolerance. *Nat Immunol* 11(1): 21-27.

Nachbur, U., Kassahn, D., Yousefi, S., Legler, D. F., Brunner, T. (2006). Posttranscriptional regulation of Fas (CD95) ligand killing activity by lipid rafts. *Blood* 107(7): 2790-2796.

Nagata, S. (1997). [Apoptosis mediated by Fas and its related diseases]. *Nippon Ika Daigaku Zasshi* 64(5): 459-462.

Nagata, S. and Suda, T. (1995). Fas and Fas ligand: lpr and gld mutations. *Immunol Today* 16(1): 39-43.

Nicoletti, I., Migliorati, G., Pagliacci, M. C., Grignani, F., Riccardi, C. (1991). A rapid and simple method for measuring thymocyte apoptosis by propidium iodide staining and flow cytometry. *J Immunol Methods* 139(2): 271-279.

O' Reilly, L., Tai, L., Lee, L., Kruse, E. A., Grabow, S., Fairlie, W. D., Haynes, N. M., Tarlinton, D. M., Zhang, J. G., Belz, G. T., Smyth, M. J., Bouillet, P., Robb, L., Strasser, A. (2009). Membrane-bound Fas ligand only is essential for Fas-induced apoptosis. *Nature* 461(7264): 659-663.

Oehme, I., Neumann, F., Bosser, S., Zornig, M. (2005). Transgenic overexpression of the Caspase-8 inhibitor FLIP(short) leads to impaired T cell proliferation and an increased memory T cell pool after staphylococcal enterotoxin B injection. *Eur J Immunol* 35(4): 1240-1249.

Orlinick, J. R., Elkon, K. B., Chao, M. V. (1997a). Separate domains of the human fas ligand dictate self-association and receptor binding. *J Biol Chem* 272(51): 32221-32229.

Orlinick, J. R., Vaishnaw, A., Elkon, K. B., Chao, M. V. (1997b). Requirement of cysteine-rich repeats of the Fas receptor for binding by the Fas ligand. *J Biol Chem* 272(46): 28889-28894.

Pao, L. I., Badour, K., Siminovitch, K. A., Neel, B. G. (2007). Nonreceptor protein-tyrosine phosphatases in immune cell signaling. *Annu Rev Immunol* 25: 473-523.

Park, S. M., Chen, L., Zhang, M., Ashton-Rickardt, P., Turner, J. R., Peter, M. E. (2010). CD95 is cytoprotective for intestinal epithelial cells in colitis. *Inflamm Bowel Dis* 16(6): 1063-1070.

Pasquale, E. B. (2008). Eph-ephrin bidirectional signaling in physiology and disease. *Cell* 133(1): 38-52.

Paulsen, M., Mathew, B., Qian, J., Lettau, M., Kabelitz, D., Janssen, O. (2009). FasL cross-linking inhibits activation of human peripheral T cells. *Int Immunol* 21(5): 587-598.

Peter, M. E., Budd, R. C., Desbarats, J., Hedrick, S. M., Hueber, A. O., Newell, M. K., Owen, L. B., Pope, R. M., Tschopp, J., Wajant, H., Wallach, D., Wiltrout, R. H., Zornig, M., Lynch, D. H. (2007). The CD95 receptor: apoptosis revisited. *Cell* 129(3): 447-450.

Qian, J., Chen, W., Lettau, M., Podda, G., Zornig, M., Kabelitz, D., Janssen, O. (2006). Regulation of FasL expression: a SH3 domain containing protein family involved in the lysosomal association of FasL. *Cell Signal* 18(8): 1327-1337.

Quintavalle, C., Incoronato, M., Puca, L., Acunzo, M., Zanca, C., Romano, G., Garofalo, M., Iaboni, M., Croce, C. M., Condorelli, G. (2010). c-FLIP(L) enhances anti-apoptotic Akt functions by modulation of Gsk3beta activity. *Cell Death Differ*.

Ramaswamy, M., Cleland, S. Y., Cruz, A. C., Siegel, R. M. (2009). Many checkpoints on the road to cell death: regulation of Fas-FasL interactions and Fas signaling in peripheral immune responses. *Results Probl Cell Differ* 49: 17-47.

Rathmell, J. C., Townsend, S. E., Xu, J. C., Flavell, R. A., Goodnow, C. C. (1996). Expansion or elimination of B cells in vivo: dual roles for CD40- and Fas (CD95)-ligands modulated by the B cell antigen receptor. *Cell* 87(2): 319-329.

Restifo, N. P. (2000). Not so Fas: Re-evaluating the mechanisms of immune privilege and tumor escape. *Nat Med* 6(5): 493-495.

Reya, T., O'Riordan, M., Okamura, R., Devaney, E., Willert, K., Nusse, R., Grosschedl, R. (2000). Wnt signaling regulates B lymphocyte proliferation through a LEF-1 dependent mechanism. *Immunity* 13(1): 15-24.

Riccardi, C. and Nicoletti, I. (2006). Analysis of apoptosis by propidium iodide staining and flow cytometry. *Nat Protoc* 1(3): 1458-1461.

Rieux-Laucat, F. (2006). Inherited and acquired death receptor defects in human Autoimmune Lymphoproliferative Syndrome. *Curr Dir Autoimmun* 9: 18-36.

Rossin, A., Kral, R., Lounnas, N., Chakrabandhu, K., Mailfert, S., Marguet, D., Hueber, A. O. (2010). Identification of a lysine-rich region of Fas as a raft nanodomain targeting signal necessary for Fas-mediated cell death. *Exp Cell Res* 316(9): 1513-1522.

Sancho-Martinez, I. and Martin-Villalba, A. (2009). Tyrosine phosphorylation and CD95: a FAScinating switch. *Cell Cycle* 8(6): 838-842.
Scaffidi, C., Fulda, S., Srinivasan, A., Friesen, C., Li, F., Tomaselli, K. J., Debatin, K. M., Krammer, P. H., Peter, M. E. (1998). Two CD95 (APO-1/Fas) signaling pathways. *EMBO J* 17(6): 1675-1687.
Scaffidi, C., Schmitz, I., Krammer, P. H., Peter, M. E. (1999). The role of c-FLIP in modulation of CD95-induced apoptosis. *J Biol Chem* 274(3): 1541-1548.
Schmidt, H., Gelhaus, C., Lucius, R., Nebendahl, M., Leippe, M., Janssen, O. (2009). Enrichment and analysis of secretory lysosomes from lymphocyte populations. *BMC Immunol* 10: 41.
Schneider, P., Holler, N., Bodmer, J. L., Hahne, M., Frei, K., Fontana, A., Tschopp, J. (1998). Conversion of membrane-bound Fas(CD95) ligand to its soluble form is associated with downregulation of its proapoptotic activity and loss of liver toxicity. *J Exp Med* 187(8): 1205-1213.
Schonwasser, D. C., Marais, R. M., Marshall, C. J., Parker, P. J. (1998). Activation of the mitogen-activated protein kinase/extracellular signal-regulated kinase pathway by conventional, novel, and atypical protein kinase C isotypes. *Mol Cell Biol* 18(2): 790-798.
Schulte, M., Reiss, K., Lettau, M., Maretzky, T., Ludwig, A., Hartmann, D., de Strooper, B., Janssen, O., Saftig, P. (2007). ADAM10 regulates FasL cell surface expression and modulates FasL-induced cytotoxicity and activation-induced cell death. *Cell Death Differ* 14(5): 1040-1049.
Schwarz, A., Grabbe, S., Grosse-Heitmeyer, K., Roters, B., Riemann, H., Luger, T. A., Trinchieri, G., Schwarz, T. (1998). Ultraviolet light-induced immune tolerance is mediated via the Fas/Fas-ligand system. *J Immunol* 160(9): 4262-4270.
Scott, F. L., Stec, B., Pop, C., Dobaczewska, M. K., Lee, J. J., Monosov, E., Robinson, H., Salvesen, G. S., Schwarzenbacher, R., Riedl, S. J. (2009). The Fas-FADD death domain complex structure unravels signalling by receptor clustering. *Nature* 457(7232): 1019-1022.
Senthilkumar, R. and Lee, H. W. (2009). CD137L- and RANKL-mediated reverse signals inhibit osteoclastogenesis and T lymphocyte proliferation. *Immunobiology* 214(2): 153-161.
Shi, Y. (2002). Mechanisms of caspase activation and inhibition during apoptosis. *Mol Cell* 9(3): 459-470.
Siegel, R. M., Frederiksen, J. K., Zacharias, D. A., Chan, F. K., Johnson, M., Lynch, D., Tsien, R. Y., Lenardo, M. J. (2000). Fas preassociation required for apoptosis signaling and dominant inhibition by pathogenic mutations. *Science* 288(5475): 2354-2357.
Siegel, R. M., Muppidi, J. R., Sarker, M., Lobito, A., Jen, M., Martin, D., Straus, S. E., Lenardo, M. J. (2004). SPOTS: signaling protein oligomeric transduction structures are early mediators of death receptor-induced apoptosis at the plasma membrane. *J Cell Biol* 167(4): 735-744.
Staal, F. J., Luis, T. C., Tiemessen, M. M. (2008). WNT signalling in the immune system: WNT is spreading its wings. *Nat Rev Immunol* 8(8): 581-593.
Stranges, P. B., Watson, J., Cooper, C. J., Choisy-Rossi, C. M., Stonebraker, A. C., Beighton, R. A., Hartig, H., Sundberg, J. P., Servick, S., Kaufmann, G., Fink, P. J., Chervonsky, A. V. (2007). Elimination of antigen-presenting cells and autoreactive T cells by Fas contributes to prevention of autoimmunity. *Immunity* 26(5): 629-641.
Strasser, A., Jost, P. J., Nagata, S. (2009). The many roles of FAS receptor signaling in the immune system. *Immunity* 30(2): 180-192.

Strater, J., Mariani, S. M., Walczak, H., Rucker, F. G., Leithauser, F., Krammer, P. H., Moller, P. (1999). CD95 ligand (CD95L) in normal human lymphoid tissues: a subset of plasma cells are prominent producers of CD95L. *Am J Pathol* 154(1): 193-201.

Streilein, J. W. (1995). Immunological non-responsiveness and acquisition of tolerance in relation to immune privilege in the eye. *Eye (Lond)* 9 (Pt 2): 236-240.

Stuart, P. M., Yin, X., Plambeck, S., Pan, F., Ferguson, T. A. (2005). The role of Fas ligand as an effector molecule in corneal graft rejection. *Eur J Immunol* 35(9): 2591-2597.

Suda, T. and Nagata, S. (1997). Why do defects in the Fas-Fas ligand system cause autoimmunity? *J Allergy Clin Immunol* 100(6 Pt 2): S97-101.

Suda, T., Takahashi, T., Golstein, P., Nagata, S. (1993). Molecular cloning and expression of the Fas ligand, a novel member of the tumor necrosis factor family. *Cell* 75(6): 1169-1178.

Sun, M., Ames, K. T., Suzuki, I., Fink, P. J. (2006). The cytoplasmic domain of Fas ligand costimulates TCR signals. *J Immunol* 177(3): 1481-1491.

Sun, M. and Fink, P. J. (2007a). A new class of reverse signaling costimulators belongs to the TNF family. *J Immunol* 179(7): 4307-4312.

Sun, M., Lee, S., Karray, S., Levi-Strauss, M., Ames, K. T., Fink, P. J. (2007b). Cutting edge: two distinct motifs within the Fas ligand tail regulate Fas ligand-mediated costimulation. *J Immunol* 179(9): 5639-5643.

Suzuki, A., Sugimura, K., Ohtsuka, K., Hasegawa, K., Suzuki, K., Ishizuka, K., Mochizuki, T., Honma, T., Narisawa, R., Asakura, H. (2000a). Fas/Fas ligand expression and characteristics of primed CD45RO+ T cells in the inflamed mucosa of ulcerative colitis. *Scand J Gastroenterol* 35(12): 1278-1283.

Suzuki, I. and Fink, P. J. (1998). Maximal proliferation of cytotoxic T lymphocytes requires reverse signaling through Fas ligand. *J Exp Med* 187(1): 123-128.

Suzuki, I., Martin, S., Boursalian, T. E., Beers, C., Fink, P. J. (2000b). Fas ligand costimulates the in vivo proliferation of CD8+ T cells. *J Immunol* 165(10): 5537-5543.

Takahashi, T., Tanaka, M., Brannan, C. I., Jenkins, N. A., Copeland, N. G., Suda, T., Nagata, S. (1994a). Generalized lymphoproliferative disease in mice, caused by a point mutation in the Fas ligand. *Cell* 76(6): 969-976.

Takahashi, T., Tanaka, M., Inazawa, J., Abe, T., Suda, T., Nagata, S. (1994b). Human Fas ligand: gene structure, chromosomal location and species specificity. *Int Immunol* 6(10): 1567-1574.

Tanaka, M., Itai, T., Adachi, M., Nagata, S. (1998). Downregulation of Fas ligand by shedding. *Nat Med* 4(1): 31-36.

Tanaka, M., Suda, T., Takahashi, T., Nagata, S. (1995). Expression of the functional soluble form of human fas ligand in activated lymphocytes. *EMBO J* 14(6): 1129-1135.

Tarlinton, D., Radbruch, A., Hiepe, F., Dorner, T. (2008). Plasma cell differentiation and survival. *Curr Opin Immunol* 20(2): 162-169.

Thornhill, P. B., Cohn, J. B., Drury, G., Stanford, W. L., Bernstein, A., Desbarats, J. (2007). A proteomic screen reveals novel Fas ligand interacting proteins within nervous system Schwann cells. *FEBS Lett* 581(23): 4455-4462.

Thornhill, P. B., Cohn, J. B., Stanford, W. L., Desbarats, J. (2008). The adaptor protein Grb2 regulates cell surface Fas ligand in Schwann cells. *Biochem Biophys Res Commun* 376(2): 341-346.

Ulisse, S., Cinque, B., Silvano, G., Rucci, N., Biordi, L., Cifone, M. G., D'Armiento, M. (2000). Erk-dependent cytosolic phospholipase A2 activity is induced by CD95 ligand cross-linking in the mouse derived Sertoli cell line TM4 and is required to trigger apoptosis in CD95 bearing cells. *Cell Death Differ* 7(10): 916-924.

Veillette, A., Latour, S., Davidson, D. (2002). Negative regulation of immunoreceptor signaling. *Annu Rev Immunol* 20: 669-707.

Veillette, A., Rhee, I., Souza, C. M., Davidson, D. (2009). PEST family phosphatases in immunity, autoimmunity, and autoinflammatory disorders. *Immunol Rev* 228(1): 312-324.

Viard-Leveugle, I., Bullani, R. R., Meda, P., Micheau, O., Limat, A., Saurat, J. H., Tschopp, J., French, L. E. (2003). Intracellular localization of keratinocyte Fas ligand explains lack of cytolytic activity under physiological conditions. *J Biol Chem* 278(18): 16183-16188.

Voss, M., Lettau, M., Janssen, O. (2009). Identification of SH3 domain interaction partners of human FasL (CD178) by phage display screening. *BMC Immunol* 10: 53.

Voss, M., Lettau, M., Paulsen, M., Janssen, O. (2008). Posttranslational regulation of Fas ligand function. *Cell Commun Signal* 6: 11.

Wajant, H. (2003). Death receptors. *Essays Biochem* 39: 53-71.

Wajant, H., Pfizenmaier, K., Scheurich, P. (2003). Non-apoptotic Fas signaling. *Cytokine Growth Factor Rev* 14(1): 53-66.

Wasem, C., Frutschi, C., Arnold, D., Vallan, C., Lin, T., Green, D. R., Mueller, C., Brunner, T. (2001). Accumulation and activation-induced release of preformed Fas (CD95) ligand during the pathogenesis of experimental graft-versus-host disease. *J Immunol* 167(5): 2936-2941.

Watts, A. D., Hunt, N. H., Wanigasekara, Y., Bloomfield, G., Wallach, D., Roufogalis, B. D., Chaudhri, G. (1999). A casein kinase I motif present in the cytoplasmic domain of members of the tumour necrosis factor ligand family is implicated in 'reverse signalling'. *EMBO J* 18(8): 2119-2126.

Weinlich, R., Brunner, T., Amarante-Mendes, G. P. (2010). Control of death receptor ligand activity by posttranslational modifications. *Cell Mol Life Sci* 67(10): 1631-1642.

Xiao, S., Deshmukh, U. S., Jodo, S., Koike, T., Sharma, R., Furusaki, A., Sung, S. S., Ju, S. T. (2004). Novel negative regulator of expression in Fas ligand (CD178) cytoplasmic tail: evidence for translational regulation and against Fas ligand retention in secretory lysosomes. *J Immunol* 173(8): 5095-5102.

Yamamoto, D., Sonoda, Y., Hasegawa, M., Funakoshi-Tago, M., Aizu-Yokota, E., Kasahara, T. (2003). FAK overexpression upregulates cyclin D3 and enhances cell proliferation via the PKC and PI3-kinase-Akt pathways. *Cell Signal* 15(6): 575-583.

Yang, H. and Reinherz, E. L. (2006). CD2BP1 modulates CD2-dependent T cell activation via linkage to protein tyrosine phosphatase (PTP)-PEST. *J Immunol* 176(10): 5898-5907.

Yu, J. W. and Shi, Y. (2008a). FLIP and the death effector domain family. *Oncogene* 27(48): 6216-6227.

Yu, Q., Quinn, W. J., 3rd, Salay, T., Crowley, J. E., Cancro, M. P., Sen, J. M. (2008b). Role of beta-catenin in B cell development and function. *J Immunol* 181(6): 3777-3783.

Zhou, Y., Gunput, R. A., Pasterkamp, R. J. (2008). Semaphorin signaling: progress made and promises ahead. *Trends Biochem Sci* 33(4): 161-170.

Zuccato, E., Blott, E. J., Holt, O., Sigismund, S., Shaw, M., Bossi, G., Griffiths, G. M. (2007). Sorting of Fas ligand to secretory lysosomes is regulated by mono-ubiquitylation and phosphorylation. *J Cell Sci* 120(Pt 1): 191-199.

6 Appendix

6.1 Systematic expression analysis of T cells

Molecular phenotyping of T cells using *Affymetrix* 21K cDNA chips: add on to 3.3.4.1 'Global gene expression profiling'.

Table A.1 Molecular phenotyping of resting T cells. Splenic T cells were isolated from *FasL ΔIntra* or wildtype mice and used for RNA preparation. The corresponding cDNA was hybridized to *Affymetrix* 21K cDNA chips for global gene expression profiling. Mean log2 ratios give the fold-change [log2^n] in *FasL ΔIntra* cells compared to wildtype cells isolated from six mice per group.

Mean log2 ratio	Array TAG ID	Gene symbol	Comment
0,88	MG-13-25j10	*Klf3*	Kruppel-like factor 3
0,71	MG-3-43i9	*Cd164*	CD164 antigen
0,71	MG-15-81c5	*Tardbp*	TAR DNA binding protein
0,68	MG-14-25e18	*AI447904*	
0,68	MG-16-190o16	*Pdcd1*	Programmed cell death 1
0,67	MG-3-8a17	*Tcp1*	T-complex protein 1
0,64	MG-3-16p4	*Odc1*	ornithine decarboxylase, structural 1
0,62	MG-15-251k6	*Tmem49*	transmembrane protein 49
0,60	MG-3-35j9	*Obfc2a*	oligonucleotide/oligosaccharide-binding fold containing 2A
0,59	MG-14-78c20	*BX634656*	
0,58	MG-14-5b16	*Dleu2*	deleted in lymphocytic leukemia, 2
0,57	MG-14-34i8	*Apg7l*	Autophagy 7-like
0,56	MG-3-264c18	*D630029K19Rik*	
0,55	MG-3-149l22	*Crem*	cAMP responsive element modulator
0,55	MG-16-49e11	*CR515106*	
0,52	MG-3-174k14	*Efcab1*	EF hand calcium binding domain 1
0,52	MG-15-92f1	*AI661017*	
0,45	MG-6-42c2	*CR520027*	
0,44	MG-3-140g13	*Polg2*	Polymerase, gamma 2, accessory subunit
-0,64	MG-15-119k2	*Jun*	Jun oncogene
-0,72	MG-16-6f22	*Ii*	Ia-associated invariant chain
-0,81	MG-14-114l18	*Myd116*	Myeloid differentiation primary response gene 116

Table A.2 Molecular phenotyping of activated T cells. Splenic T cells were isolated from *FasL ΔIntra* or wildtype mice, stimulated for four hours *ex vivo* (1 µg/ml plate-bound anti-CD3 antibody) and used for RNA preparation. The corresponding cDNA was hybridized to *Affymetrix* 21K cDNA chips for global gene expression profiling. Mean log2 ratios give the fold-change [$\log 2^n$] in *FasL ΔIntra* cells compared to wildtype cells isolated from six mice per group.

Mean log2 ratio	Array TAG ID	Gene symbol	Comment
2,17	MG-12-205b11	*1500005K14Rik*	
2,04	MG-15-194f15	*Slc28a2*	Solute carrier family 28, member 2
1,89	MG-6-64e20	*2610019F03Rik*	
1,83	MG-6-68f24	*Camk4*	Calcium/calmodulin-dependent protein kinase IV
-1,20	MG-15-186e9	*Pitrm1*	Pitrilysin metalloprotease 1
-1,22	MG-16-170o8	*Npm1*	Nucleophosmin 1
-1,23	MG-3-96p15	*Wdr36*	WD repeat domain 36
-1,23	MG-26-133l1	*Eif3s9*	Eukaryotic translation initiation factor 3, subunit 9
-1,25	MG-8-23g17	*Gcs1*	Glucosidase 1
-1,25	MG-14-102l11	*Mrps18b*	Mitochondrial ribosomal protein S18B
-1,25	MG-16-134i15	*Sfxn1*	Sideroflexin 1
-1,26	MG-3-276j1	*Mpp6*	Membrane protein, palmitoylated 6
-1,26	MG-3-26p20	*8430427H17Rik*	
-1,26	MG-3-99i22	*Rnu22*	RNA, U22 small nucleolar
-1,26	MG-16-5g10	*BX634198*	
-1,27	MG-16-159g21	*Cacybp*	Calcyclin binding protein
-1,27	MG-3-277p13	*Bxdc2*	brix domain containing 2
-1,28	MG-14-20i15	*Gtpbp4*	GTP binding protein 4
-1,28	MG-3-24h9	*1700129C05Rik*	
-1,29	MG-16-9e13	*Ftsj3*	FtsJ homolog 3
-1,29	MG-16-52c2	*CR516105*	
-1,30	MG-15-106c11	*3110082I17Rik*	
-1,30	MG-3-54n14	*Aatf*	Apoptosis antagonizing transcription factor
-1,31	MG-12-248g10	*D15Ertd785e*	DNA segment, Chr 15, ERATO Doi 785, expressed
-1,31	MG-16-10m22	*6720458F09Rik*	
-1,31	MG-3-21h7	*A330048O09Rik*	

		k	
-1,32	MG-3-62o14	Jtv1	JTV1 gene
-1,33	MG-6-43d8	Ak2	Adenylate kinase 2
-1,33	MG-16-6d20	Rars	Arginyl-tRNA synthetase
-1,33	MG-12-132a20	Ddx18	DEAD (Asp-Glu-Ala-Asp) box polypeptide 18
-1,34	MG-15-52g13	Hrb2	HIV-1 Rev binding protein 2
-1,34	MG-14-72k7	CR514785	
-1,34	MG-8-49h4	Shmt2	Serine hydroxymethyl transferase 2
-1,35	MG-4-5j23	Nfkb1	nuclear factor of kappa light polypeptide gene enhancer in B-cells 2
-1,35	MG-3-45l15	Sptlc1	Serine palmitoyltransferase, long chain base subunit 1
-1,36	MG-26-213i9	Raver1	ribonucleoprotein, PTB-binding 1
-1,36	MG-8-40d18	CR517526	
-1,37	s1-B130016L12Rik	Heatr1	HEAT repeat containing 1
-1,38	MG-15-224e21	BX635857	
-1,38	MG-14-80g20	Ddx3x	Fibroblast growth factor inducible 14
-1,38	MG-6-1a17	Cycs	Cytochrome c, somatic
-1,38	MG-3-38d12	Utp6	UTP6, small subunit processome component
-1,38	MG-15-25d23	Xpot	Exportin, tRNA
-1,39	MG-3-105k23	Hspca	Heat shock protein 1, alpha
-1,39	MG-6-54e3	Acsl5	Acyl-CoA synthetase long-chain family member 5
-1,41	MG-8-26m3	BX636174	
-1,41	MG-3-23e8	D130017N08Rik	
-1,42	MG-3-73i12	2610528E23Rik	
-1,42	MG-16-3e3	2410015N17Rik	
-1,43	MG-26-111f17	Shmt2	Serine hydroxymethyl transferase 2
-1,43	MG-6-40l14	Gart	Phosphoribosylglycinamide formyltransferase
-1,44	MG-14-117n21	D8Ertd319e	DNA segment, Chr 8, ERATO Doi 319, expressed
-1,44	MG-16-4n14	Lrmp	Lymphoid-restricted membrane protein
-1,44	MG-15-232l20	BC065123	
-1,44	MG-14-107k7	Nfatc1	Nuclear factor of activated T-cells

-1,44	MG-6-64c5	Rnf149	Ring finger protein 149
-1,45	MG-6-13k9	Slc6a17	Solute carrier family 6 member 17
-1,45	MG-6-92e18	Slc25a33	solute carrier family 25, member 33
-1,45	MG-6-81e17		
-1,45	MG-6-63c21	Tars	Threonyl-tRNA synthetase
-1,45	MG-16-152i24	5830405N20Rik	
-1,46	Nomo1	Nomo1	nodal modulator 1
-1,47	MG-16-273g23	Kti12	KTI12 homolog, chromatin associated
-1,47	MG-14-83j4	Ldha	lactate dehydrogenase A
-1,48	MG-15-106g12	Ccdc115	coiled-coil domain containing 115
-1,49	MG-26-276i12	Rsl1d1	ribosomal L1 domain containing 1
-1,49	MG-3-37k5	Mrpl12	Mitochondrial ribosomal protein L12
-1,49	MG-6-38f21	Timm10	Translocase of inner mitochondrial membrane 10 homolog
-1,49	MG-15-144h5	Numb	Numb gene homolog
-1,49	MG-15-172d6	Ppat	Phosphoribosyl pyrophosphate amidotransferase
-1,52	MG-3-75c15	4921511H13Rik	
-1,52	MG-3-26n8	Pomp	proteasome maturation protein
-1,53	MG-8-60f17	Nola2	Nucleolar protein family A, member 2
-1,54	MG-3-35p10	Rnu3ip2	RNA, U3 small nucleolar interacting protein 2
-1,55	MG-15-1m17	Acsl5	Acyl-CoA synthetase long-chain family member 5
-1,55	MG-12-242l23	Pphln1	Periphilin 1
-1,55	MG-14-81c3	AY078069	
-1,57	MG-6-39f17	Abcf1	ATP-binding cassette, sub-family F
-1,58	MG-6-31p20	CR521724	
-1,58	s0-Ccl4	Ccl4	chemokine (C-C motif) ligand 4
-1,59	MG-8-57c8	2410015N17Rik	
-1,59	MG-8-118p15	D5Ertd606e	DNA segment, Chr 5, ERATO Doi 606, expressed
-1,59	MG-8-98a20	Umps	Uridine monophosphate

			synthetase
-1,59	MG-16-52e10	Sla	Src-like adaptor
-1,60	MG-3-4n16	Bnc2	basonuclin 2
-1,61	MG-15-255e11	Tacc2	Transforming, acidic coiled-coil containing protein 2
-1,61	MG-12-133j12	Sema7a	Sema domain, immunoglobulin domain, 7A
-1,61	MG-15-260i19	Txnrd1	thioredoxin reductase 1
-1,61	MG-68-88b2	Chc1	Chromosome condensation 1
-1,62	MG-3-57f5	Hspa9	heat shock protein 9
-1,62	MG-6-46j8	Polr2h	Similar to polymerase (RNA) II polypeptide H
-1,62	MG-15-188g13	BX633985	
-1,66	MG-15-74l4	Irf4	Interferon regulatory factor 4
-1,67	MG-3-56e17	Pprc1	Peroxisome proliferative activated receptor
-1,67	MG-3-75g7	Lyar	Ly1 antibody reactive clone
-1,68	MG-3-106h2	Hspa9	heat shock protein 9
-1,68	MG-3-31j8	Tmem97	transmembrane protein 97
-1,69	MG-15-271k2	Mthfd2	Methylenetetrahydrofolate dehydrogenase
-1,69	MG-26-261n18	Rrp1b	ribosomal RNA processing 1 homolog B
-1,70	MG-16-40e1	Kpnb3	Karyopherin (importin) beta 3
-1,71	MG-15-120k7	BX638845	
-1,72	MG-15-71l6	Wdr75	WD repeat domain 75
-1,72	MG-3-38j18	Mrps6	Mitochondrial ribosomal protein S6
-1,72	MG-15-3c13	Tagap1	T-cell activation GTPase activating protein 1
-1,74	s1-Mrps6	Mrps6	mitochondrial ribosomal protein S6
-1,74	MG-16-45k7	5830405N20Rik	
-1,75	MG-6-82j10	Smyd5	SET and MYND domain containing 5
-1,75	MG-3-81o13	Mrto4	MRT4, mRNA turnover 4, homolog
-1,75	MG-15-82f10	Wrd43	WD repeat domain 43
-1,76	MG-12-244i21	Dnmt3a	DNA methyltransferase 3A
-1,76	MG-26-78j20	Cad	Carbamoyl-phosphate synthetase 2
-1,76	MG-8-20p3	Nol1	Nucleolar protein 1
-1,77	s0-5830438K24Rik	Cd59	CD69 antigen
-1,78	MG-16-183g4	BC018399	

-1,79	MG-13-41c7	*Galk1*	Galactokinase 1
-1,79	MG-12-269j19	*BX634314*	
-1,79	MG-26-239b21	*CR518583*	
-1,80	MG-3-166n5	*Mdn1*	Midasin homolog
-1,80	MG-3-123d23	*Pwp2h*	PWP2 homolog, yeast
-1,81	MG-6-19l12	*Eif5a*	Eukaryotic translation initiation factor 5A
-1,81	MG-15-146i23	*Gpatch4*	G patch domain containing 4
-1,82	MG-15-260m2	*Sdad1*	SDA1 domain containing 1
-1,83	MG-68-175p16	*Timm13a*	Translocase of inner mitochondrial membrane 13 homolog a
-1,84	MG-3-24h16	*Fbl*	Fibrillarin
-1,85	MG-3-101d3	*CR515925*	
-1,85	MG-6-77k11	*Ctps*	Cytidine 5--triphosphate synthase
-1,85	MG-13-57n5	*Nol5a*	Nucleolar protein 5A
-1,85	MG-16-44g1	*BX637222*	
-1,86	MG-13-1a19	*AI429152*	
-1,86	MG-15-2a10	*2310056P07Rik*	
-1,87	MG-3-46a10	*Slamf7*	SLAM family member 7
-1,87	MG-68-170k10	*Ctps*	Cytidine 5--triphosphate synthase
-1,89	MG-15-210g10	*5830428M24Rik*	
-1,89	MG-26-118g15	*Nsun2*	NOL1/NOP2/Sun domain family 2
-1,90	MG-8-19b7	*Fabp5*	Fatty acid binding protein 5, epidermal
-1,93	MG-15-36o13	*Hdac5*	Histone deacetylase 5
-1,95	MG-15-239g10	*Bxdc1*	Brix domain containing 1
-1,97	MG-8-30k19	*Mki67ip*	Mki67 interacting nucleolar phosphoprotein
-1,97	MG-3-12a18	*Ipo4*	Importin 4
-2,05	MG-14-49d18	*C1qbp*	Complement component 1, q subcomponent binding protein
-2,05	MG-12-1g17	*Atad3a*	ATPase family, AAA domain containing 3A
-2,07	MG-3-40e12	*C230062I16Rik*	
-2,08	MG-16-134n7	*Fubp1*	Far upstream element binding protein 1
-2,10	MG-68-113c23	*Tnf*	Tumor necrosis factor
-2,13	MG-12-234j10	*4933433P14Rik*	
-2,15	MG-15-20g11	*Slc1a5*	Solute carrier family 1,

			member 5
-2,16	MG-15-164g13	*BX632851*	
-2,17	MG-3-88e11	*C77032*	
-2,17	MG-16-3j5	*Timm8a*	Translocase of inner mitochondrial membrane 8 homolog a
-2,17	MG-13-133o15	*BX632490*	
-2,21	MG-3-256a12	*Ipo4*	Importin 4
-2,22	MG-12-281j21	*Ifrd2*	Interferon-related developmental regulator 2
-2,25	MG-3-84p19	*CR519854*	
-2,28	MG-6-31f19	*Gramd1b*	GRAM domain containing 1B
-2,28	MG-16-41i17	*Slc29a1*	Solute carrier family 29 member 1
-2,31	MG-6-65p24	*D1Ertd471e*	DNA segment, Chr 1, ERATO Doi 471, expressed
-2,33	MG-3-228n12	*AA408556*	
-2,33	s0-Cxcl10	*Cxcl10*	chemokine (C-X-C motif) ligand 10
-2,34	MG-68-147b5	*Ubash3b*	ubiquitin associated and SH3 domain containing, B
-2,37	MG-6-42b20	*Ccnd2*	Cyclin D2
-2,37	MG-16-154p22	*AF155546*	
-2,39	MG-13-58k4	*Nola1*	Nucleolar protein family A, member 1
-2,41	MG-68-182c20	*Chi3l1*	Chitinase 3-like 1
-2,41	MG-3-8f14	*Nol5*	Nucleolar protein 5
-2,41	MG-3-33g7	*Slc7a1*	Solute carrier family 7, member 1
-2,43	MG-6-41c3	*Mthfd1*	Methylenetetrahydrofolate dehydrogenase
-2,43	MG-15-151a5	*Nr4a1*	Nuclear receptor subfamily 4, group A, member 1
-2,45	MG-6-24n11	*AI850995*	
-2,53	MG-3-138c23	*Gramd1b*	GRAM domain containing 1B
-2,54	MG-3-56n13	*Slc7a5*	Solute carrier family 7, member 5
-2,55	MG-6-42c2	*CR520027*	
-2,57	MG-4-145m17	*Srm*	Spermidine synthase
-2,57	MG-4-145h9	*Serpina3g*	Serine proteinase inhibitor, clade A, member 3G
-2,58	MG-26-241h16	*Srm*	Spermidine synthase
-2,65	MG-16-264l1	*Apol7c*	apolipoprotein L 7c
-2,78	s0-Ccl3	*Ccl3*	chemokine (C-C motif) ligand 3
-2,81	MG-73-18e3	*Nr4a2*	Nuclear receptor subfamily 4, group A, member 2

| -2,95 | s0-Egr1 | *Egr1* | early growth response 1 |
| -3,31 | MG-14-5l12 | *Chst2* | Carbohydrate sulfotransferase 2 |

6.2 Abbreviations

°C	degree Celsius
µg	microgram
µl	microliter
µM	micromolar
µm	micrometer
aa	amino acid
AAD	allergic airway disease
ACAD	activated cell autonomous death
ADAM	A disintegrin and metalloprotease
AICD	activation-induced cell death
ALPS	autoimmune lymphoproliferativ syndrome
alum	aluminium-potassium-dodecahydrate (KAL(SO$_4$)$_2$)
AP	Alcaline phosphatase
AP-1	Activator protein-1
Apaf-1	Apoptotic protease activating factor-1
APC	antigen-presenting cell
APS	ammoniumperoxodisulphate
ATCC	American type culture collection
ATP	adenosine-5´ triphosphate
BAL	bronchoalveolar lavage
Bcl-2	B-cell CLL/lymphoma associated-2
BCR	B cell receptor
BH3-only	Bcl-2 homology domain 3-only
BHI agar	brain heart infusion agar
Blimp-1	B lymphocyte-induced maturation protein 1
bp	base pair(s)
BrdU	bromo desoxyuridine
BSA	Bovine serum albumine
Btk	Bruton's tyrosine kinase
C	cystein
C57BL/6	C57 black 6 incest mouse
Cas	Breast cancer antiestrogen 1; BCAR 1
Casp-	Caspase
CD	cluster of differentiation
CD2BP1	CD2-binding protein-1
CDK	Cyclin-dependent kinase
cDNA	complementary DNA
c-FLIP	Cellular FLICE-like inhibitory protein
CFSE	Carboxyfluorescein-diacetat-succinimidylester
CFU	colony forming units
CIITA	MHC class II transactivator
CK-I	Casein kinase I
cm	centimeter
CMV	cytomegalovirus
cPLA$_2$	Cytosolic phospholipase A2

CRD	Cystein-rich domain
Csk	Cytoplasmic tyrosine kinase
C_t	threshold cycle
CTL	cytotoxic T cell
CTLA-4	Cytotoxic T-lymphocyte associated protein-4
Cyt c	Cytochrom c
DAG	diacylglycerole
dATP	2´-deoxy adenosine-5 triphosphate
DC	dendritic cell
DcR	decoy receptor
DD	death domain
DDM	doublet discrimination module
DED	death effector domain
DISC	death-inducing signaling complex
DMEM	Dulbecco's modified Eagle's medium
DMSO	dimethyl sulphoxide
DNA	deoxyribonucleic acid
dNTP	desoxy nucleotide triphosphate
DTT	dithiothreitol
ECD	extracellular domain
ECL	enhanced chemoluminescence light
EDA	Ectodysplasin A
EDTA	ethylenediaminetetraacetic acid
e.g.	exempli gratia (for example)
Egr	Early growth response
ELISA	enzyme-linked immuno absorbent assay
ERK	Extracellular signal-regulated kinase
EtOH	ethanol
FACS	fluorescence-activated cell sorting
FADD	Fas-associated protein with death domain
FasL	Fas Ligand
Fc	fragment crystallizable region
FCS	fetal calve serum
FGF	Fibroblast growth factor
FITC	fluoresceinisothiocyanat
FL	fluorescence channel
FL2-A	fluorescence channel 2-area
FL2-W	fluorescence channel 2-width
FoB	follicular/ transitional B cell
FSC	forward scatter
g	gravity
G	guanine
Gads	Grb2-related adaptor downstream of Shc
GAPDH	Glycerinaldehyde-3-phosphate dehydrogenase
GC	germinal center B cell
gDNA	genomic DNA
gld	generalized lymphoproliferative disease
GMC	German Mouse Clinic, Munich
Grb-2	Growth factor binding protein 2

h	hour(s)
HIV	human immunodeficiency virus
HPLC	high pressure liquid chromatography
HPRT1	Hypoxanthine phosphoribosyltransferase 1
HRP	Horseradish peroxidase
i.p.	intraperitoneal
i.v.	intravenous
IAP	Inhibitor of apoptotsis
ICAD	Inhibitor of caspase activated DNase
ICD	intracellular domain
IFN-γ	Interferon-gamma
Ig	immunglobuline
IκB	Inhibitor of NF-κB
IL	Interleukine
IP3	Inositol (1,4,5)-trisphosphate
IRF	Interferon regulatory factor
ITAM	immunoreceptor tyrosine activation motif
ITIM	immunoreceptor tyrosine inhibitory motif
Itk	IL-2-inducible T cell kinase
JNK	c-Jun N-terminal kinase
kb	kilobase
kDa	kilodalton
kHz	kilohertz
L	lysine
l	liter
LAT	Linker for activation of T cells
Lck	Lymphocyte-specific protein tyrosine kinase
LCMV	lymphocytic choriomeningitis virus
Lef-1	Lymphoid enhancer binding factor 1
LIGHT	lymphotoxin-like, exhibits inducible expression and competes with herpes simplex virus glycoprotein D for herpes virus entry mediator, a receptor expressed by T lymphocytes
lpr	lymphoproliferation
LTαβR	Lymphotoxin receptor alpha, beta
M	molar
MΦ	macrophage
mA	milli Ampere
MACS	magnetic-activated cell sorting
MAPK	Mitogen-activated protein kinase
MEK	Mitogen-activated ERK-activating kinase
MFI	mean fluorescence intensity
mg	milligram
MHC	Major histocompatibility complex
min	minute(s)
ml	milliliter
mM	millimolar
MOMP	mitochondrial outer membrane permeabilization
mRNA	messenger ribonucleic acid

MW	molecular weight
MZB	marginal zone B cell
NaOAc	sodium acetate
NFAT	Nuclear factor of activated T cells
NF-κB	Nuclear factor-κB
NK	natural killer cells
nm	nanometer
nM	nanomolar
NP-40	Nonidet P-40
NP-CGG	4-hydroxy-3-nitrophenyl acetate chicken gamma globuline
o/n	overnight
OD	optical density
OVA	Ovalbumin
P/S	penicillin/streptomycin
PAGE	polyacrylamide gel electrophoresis
PARP	Poly (ADP-ribose) polymerase
PBMC	peripheral blood mononuclear cells
PBS	phosphate buffered saline
PC	plasma cell
PCD	programmed cell death
PCH	pombe cdc15 homology
PCR	polymerase chain reaction
PE	phycoerythrin
PEI	polyethyleneimine
PFU	plaque forming units
pg	picogram
PH	plecstrin homology
pH	pondus Hydrogenii
PI	propidium iodide
PI3K	Phosphoinositide 3 kinase
PIP2	phosphatidyl-inositol 4,5-bisphosphat
PKB	Proteinkinase B (Akt)
PKC	Proteinkinase C
PLAD	preligand assembly domain
PLCg	Phospholipase C gamma
PMA	phorbol 12-myristat-13-acetat
PMSF	phenyl-methyl-sulfonylfluoride
PRD	proline-rich domain
PST-PIP	Proline, serine, threonine phosphatase interacting protein
PTP-PEST	Protein tyrosine phosphatase-PEST, non receptor type 12
Pyk 2	Proline-rich tyrosine kinase 2
qRT-PCR	quantitative real-rime PCR
RFU	relative fluorescence unit
RICD	restimulation-induced cell death
RIP	receptor-interacting protein
RMPI	Rosewell park memorial
RNA	ribonucleic acid
RNase	Ribonuclease
rpm	rounds per minute

RT-PCR	reverse transcriptase-PCR
RU	relative units
s	second(s)
S	serine
SEB	*Staphylococcus* enterotoxin B
SEM	standard error of the mean
SDS	sodium dodecyl sulphate
SDS-PAGE	SDS-polyacrylamide-gel electrophoresis
sFasL	soluble Fas Ligand
SFK	Src family kinases
SH2/3	Src homology 2/3
SHIP	SH2 domain-containing inositol polyphosphate 5'phosphatase
SHP1/2	Protein tyrosine phosphatase 1/2
SP-1	Stimulating protein 1
SPOT	signaling protein oligomerization transduction structure
SPPL2a	Signal peptidase-like 2a
SSC	side scatter
T	threonine
TAE	tris-acetate-EDTA buffer
TBE	tris-borate-EDTA buffer
tBid	truncated Bid
TCR	T cell receptor
TEMED	N,N,N'N'-tetramethylethylendiamin
TGF	Transforming growth factor
THD	TNF-homology domain
TM	transmembrane domain
TNF	Tumor necrosis factor
TNFR	TNF receptor
TRADD	TNFR-associated protein with a death domain
TRAF	TNFR-associated factor
TRAIL	TNF-related apoptosis-inducing ligand
TRANCE	TNF-related activation-induced cytokine receptor
Tris	tris(hydroxymethyl)aminomethane
Tween 20	polyoxyethylenesorbitan monolaurate
U	units
UTR	untranslated region
UV	ultraviolett
V	volt
v.s.	versus
v/v	volume per volume
w/v	weight per volume
WASP	Wiskott-Aldrich syndrome protein
WB	Western blot
WT	wildtype
WW	protein domain containing two conserved tryptophane residues
X	any amino acid
Xbp-1	X box-binding protein-1
Y	tyrosine
ZAP-70	ζ-chain-associated protein kinase of 70 kDa

I want morebooks!

Buy your books fast and straightforward online - at one of world's fastest growing online book stores! Environmentally sound due to Print-on-Demand technologies.

Buy your books online at
www.morebooks.shop

Kaufen Sie Ihre Bücher schnell und unkompliziert online – auf einer der am schnellsten wachsenden Buchhandelsplattformen weltweit! Dank Print-On-Demand umwelt- und ressourcenschonend produziert.

Bücher schneller online kaufen
www.morebooks.shop

KS OmniScriptum Publishing
Brivibas gatve 197
LV-1039 Riga, Latvia
Telefax:+371 686 204 55

info@omniscriptum.com
www.omniscriptum.com

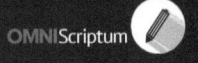

Printed by Books on Demand GmbH, Norderstedt / Germany